GrandDesigns Handbook

GRAND DESIGNS
HANDBOOK

KEVIN McCLOUD

The blueprint for building
your dream home

Collins

This paperback edition published in 2009 by Collins

First published in hardback in 2006
by Collins, an imprint of
HarperCollinsPublishers
77–85 Fulham Palace Road
Hammersmith
London W6 8JB

www.harpercollins.co.uk

Collins is a registered trademark of HarperCollins Publishers Ltd

014	013	012	011	010	09
6	5	4	3	2	1

Technical text contributors: George Baxter, Roy Speer

Editor: Zia Mattocks
Art Direction: Mark Thomson
Design: Christopher Wilson, Annabel Rooker

A catalogue record for this book is available
from the British Library

ISBN 978-0-00-730742-5

Colour reproduction by Colourscan, Singapore
Printed and bound in the UK by Butler Tanner & Dennis Ltd

ter ferns *
able into TLC *
dress. *
ds. (Fri) — seal!
lights *
hool
runs.
@ ferns leisureco
jonseating
el + cot
ides → sofa
Justo
curvy
Art
TV

bike racks | mirror | torch
table tools all / cover table.
= Filler
= Black Mastic — wetroom door.
POD. (STAINLESS) ON UPPER S/CASE.
0033 32181+563
Martin Harman 1-3
* Called:-
BOB
914 TOOK FR
4089 TUESDAY
Jeremy 695562
7—7.30
O CHEQUE + DISCLAIMER
FLAT ESTATE
BORDER
O 22 Swel 8 panels
T T S
SAI LA TAVR 80/8

MON T
£40

gD COUNTDOWN
M | T | W | T | F | S | S
4 cropper | Mode/ | Mode | BEN DAG(!)
20 | 27 | 28 Tractor
HI Ti
Floor
Boo
Julie
ANNA

Contents

I think there's no better phrase to sum up what architecture should do than 'make you feel like a better human being'. It's not a faultlessly brilliant line and I have a sneaking suspicion that I may have nicked it from someone else. But it is concise: it does sum up a great deal about our built environment and what it should do to us. Max Fordham, the famous building engineer and one of the Stirling Prize judges in 2005, evinces the same idea more personally. When he visits a building, he's always looking for an 'uplifting experience', some positive effect on his own sense of wellbeing. I'm all for that. But so few buildings provide that experience that it makes you wonder how easy that effect is to achieve.

So what are the magic ingredients in great architecture? What makes a brilliant piece of design? What are the salient features that make a house a civilized, stimulating place to live? The truth is, of course, that great buildings are born of passion, talent, a committed client, inventiveness and a sense of play, a respect for everyday ergonomics, an understanding of universal human needs and desires, and also of cultural ideas, a good budget, a helpful planning system, fine engineering and a thorough design regime. Not much then. You can begin to see why architects train for up to seven years and even then don't usually produce anything of any worth until they're in their forties.

previous page: **Bruno and Denise Del Tufo's water tower conversion in Ashford, Kent.**

Whether you build a house yourself or whether you commission one, the risks are just as great and the voyage just as epic: it isn't something anyone takes on lightly. Despite the big demographic upheavals that are occurring at the moment, our society is still structured around the family unit, each unit belonging in one home. That's a generality, but it's also an under-riding driving force in our culture. We form very powerful and primitive attachments to our nests: they make us feel safe, comfortable, happy, secure and relaxed. So if we decide to scatter them to the wind and look to start again, that means big personal upheaval. The ambition and bravery needed to do that, particularly if you have children, makes the experience akin in scale to trying to sail around the world or trek to the North Pole. That may sound an exaggeration, but the emotional and physical demands made on people going through the big change of building a bespoke house are comparable. You uproot yourselves in a similar way and you embrace change in a big way. As a result, you often become entirely different people: enriched (emotionally), impoverished (financially), wiser and, er, much, much more tired.

It's exactly this heroic and poetic quality in the people who do it that makes *Grand Designs* so watchable on television. Not the things that go wrong, or my carpings, but the underlying redemptive nature of the changes people go through. It is proper old-fashioned storytelling.

That apart, I'm aware that when we pick up the stories, people have already got their plot, their planning permission and, very often, their builder. Their architect has been on board for months, if not years, and so we're never able to cover satisfactorily the beginnings of the great adventure, nor chart all the early decision-making. This book is a guide to that stormy early period, an answer to all those questions: How do I find a good architect? Or builder? How do I navigate planning permission or listed building consent? What are building regulations? How do I figure out what I want? What colour hard hat should I buy? (Answer: green. It says so much about you.)

Having spent nearly eight years filming *Grand Designs*, I felt it was about time they let me off my leash and I gave you the benefit of my untrammelled opinion. For me, this book is about doing it right, getting through the process

This book is a guide to that stormy early period, an answer to all those questions.

with the minimum of fuss and maximum of enjoyment. Consequently, it's fairly opinionated in places. But it's also a book about architecture and how we should be raising the bar both in terms of how well designed our buildings are and how well built. In 2005 the Commission for Architecture and the Built Environment (CABE) produced a report that gave Britain's new housing an almost terminally ill bill of health. It said that just 17 per cent of new houses were any good and the rest were either 'mediocre' or 'poor'. Its Chief Executive, Richard Simmons, said that the vast majority of houses in Britain were 'not very nice places to be'. Another CABE report on housing in the North showed the situation to be even worse there, with a whopping 24 per cent of houses judged 'poor' and an almost insignificant six per cent judged either 'good' or 'very good'.

Houses, too, can be architecture, as well as the Scottish Parliament or the Gherkin.

Which gets my goat. Because if *Grand Designs* does anything for the 99 per cent of its four million viewers, it shows them that our homes can be genuinely life affirming, that houses, too, can be architecture, as well as the Scottish Parliament or the Gherkin. And it shows them what's possible, not just at the top of the market, but what might be obtainable right across Britain's housing. It raises the bar.

We need to improve the performance of our buildings as well. The building regulations are, by dint of being government policy, a compromise between the interests of hundreds of differing lobbying agencies. And as a result of being a compromise they represent a baseline, a minimum standard with which bog-ordinary developments have to comply. I'm consequently never impressed when I meet a self-builder whose project barely scrapes through the regulations. Our one-off bespoke homes in this country ought to be wonderful examples of construction. If our homes are pieces of engineering, they should be efficient pieces of engineering. We should be super-insulating our homes, minimizing their environmental impact and considering their whole lifecycle efficiency. And we should be doing so far in excess of what government regulations require us to do. That way, we not only build a house that is modern now, we build one that will still be modern in 20 years.

There are a lot of lists in this book, I admit. Mainly checklists but a few finger-wagging proscriptive ones as well. And so to finish, I thought I'd include one here, too – a list of the top ten dos and don'ts, the list that every magazine journalist asks me for. The list that, even if you close this book for ever right now and give it to the Dorothy House charity shop, you should cut out and keep.

1 Employ an architect – and a good one at that – who shares your view of the world. Don't try to design your home yourself; it's not what you were trained to do, nor will you be any good at it. Look, I've had all sorts of design training and I would still employ an architect, and on a full-service contract at that. A good architect is free (all is explained, see page 44) and will give you back your dream fully wrought and magnificently detailed. Provided that you choose the right one.

2 Contain your budget and contain your ambition, allowing yourself up to 20 per cent contingency fee on the project. If you don't spend it on the house, you can spend it on sofas and a holiday at the end. You'll need it.

3 Spend three years planning your project and putting everything in place in advance, and chances are you'll be on site six to eight months (which is the very expensive bit). Spend six months planning and you're likely to be on site for three years (which, to remind you, is the very expensive bit).

4 Don't risk all for a dream you've never lived. Don't be beguiled by all you see in magazines. Instead, incorporate as many elements in your design that you know for sure make you happy.

5 Employ as many professionals as you can afford, especially a good project manager or builder. Getting the build costed by a professional quantity surveyor is also absolutely necessary. Never rely on a costing produced by your builder or your architect: the entire construction industry is populated by optimists.

6 Don't throw away all your old furniture and buy everything entirely new. Our possessions are our autobiographies.

7 Don't expect the finished house to change who you are. It's more likely that the process of building it will teach you more about yourself.

8 Don't expect too much of your team. There is an ancient rule of thumb that can be applied to any made thing, be it a spoon or a house, which involves three variables: time, cost and quality. Put briefly, if you want the very best quality, expect to have to wait and pay for it. If you want something cheap, expect that it'll be shoddy and still take a long time. (You get the picture: change one of these variables and you change the other two). Moreover, you're very unlikely to get your own way with more than one aspect. Few people are lucky enough to get the very best quality quickly, albeit still paying through the nose. Fewer still will get a cheap house quickly and no one I've met has ever got cheap craftsmanship in double-quick time. If you meet someone who says they have, they're lying.

9 Do write everything down. Everything. Keep records, delivery notes and invoices, and ensure that you have signed contracts with all the players involved. And keep a build diary to record progress and bore your grandchildren with.

10 Do remember that building a house is not a sophisticated piece of DIY, nor is it an extended piece of interior design. The inside of your home is private and your personal realm of expression. The outside is public property and part of other people's lives. Architecture has a public responsibility and your home will become part of a landscape or townscape, so stand back from the design process and give your architect, landscape designer and planner room for manoeuvre here. As the French planner and building colourist Jean-Philippe Lenclos said, 'If you want to paint your front door, get your neighbour to choose the colour.'

1 Thinking

I've just given you a short list of all the things that can contribute to making a great building. The most important contributor is you, of course. You hold the purse strings, you choose the architect, you are the client. And yet so many people feel somehow unempowered, not empowered, by the experience of commissioning a house, which isn't how it should be. In my view, that's because although they think they know what they want, like taps and flooring and a view of one particular tree, in truth they're usually coming to the experience for the first time with no understanding of the bigger picture or even of what architecture can give them. The nearest thing they'll have experienced is ordering a bespoke kitchen from Magnet or having a suit made. Or going to the hairdresser's. Of all these, the hairdresser's can be the most terrifying because of the way you place your appearance and how you project yourself entirely in the hands of someone else. The architect is like a hairdresser, except that he or she is going to deliver you possibly 20 or 30 years of your future. They're not going to determine how you look, but how you live.

previous page: **Anjana and David Devoy's curved house in Clapham, London.**
facing page: **Cedar House, in Norfolk, is a simple, understated building that maximizes the opportunities for outdoor living.**

So it's important first to understand what you can get out of architecture, how high you can set the benchmark, before you even begin to formulate your brief to your architect.

What makes architecture?

I use four definitions of what good architecture is, according to the mood I'm in and according to how bored I think my conversational partner might get. They increase in complexity but the **first** is my favourite because it's quick and grabs you by the heart. It's one I've used for years and the one used in the introduction. It's one idea. Great architecture makes you feel like a better human being. That's its role in one simple statement, and you can use that phrase with alacrity in your brief to your architect, as far as I'm concerned. The effect of great architecture is effortless, requiring you to do nothing more than inhabit a building to benefit from it. So I recommend you demand to be made to feel like a better human being and change architects immediately if you suspect that you will be asked to make compromises and adapt your way of life to some kind of regime that the building is going to impose. That will result in the architecture making you feel like a worse human being.

Great architecture makes you feel like a better human being. That's its role in one simple statement.

The **second** definition uses two ideas – that of people and place. In other words, what separates any old building from a great building is: 1) the way in which proper architecture is designed around the people who commission it and use and inhabit it everyday – that is, you and the way you live your lives; and 2) the way in which proper architecture responds to place, to the uniqueness of its setting, so that it doesn't look as though it could belong anywhere else. This is a very useful definition because at a stroke it helps you to understand why noddy houses all over the land and out-of-town grey superstores like Halfords look out of place and odd. In being designed to fit anywhere, they fit nowhere.

In complete contrast to this method of designing buildings is an approach adopted by the architects of Howth House in Dublin (see page 141 and 152–3). John Tuomey and Sheila O'Donnell are interested in the way that landscape and setting can inspire buildings, lyrically and poetically. They visited the site at the start of the project, sat on the grass to enjoy the view and had a picnic, and from that experience and in partnership with their client, the design evolved. Similarly, I know of one architect who visited the site nine times and spent six months talking to his clients and producing nothing but flow diagrams of how they used their domestic spaces before he came up with so much as a sketch proposal. What was interesting was that the sketch proposal stunned his clients. They said that it was as though he had reached inside their minds and hearts and produced what they dared not even dream about. There's lyricism.

The **third** definition uses three ideas and it's the one that most architects will feel most at home with, being neither lyrical nor effortless. It's a definition based on the writings of Vitruvius, that architecture should provide 'firmness, commodity and delight'. This needs some explaining. Firmness can generally be understood to mean that the house shouldn't fall down or disintegrate. Commodity is another word for comfort, here taken to mean ergonomic comfort, such as doors made to the right height with door handles at hand level – that sort of obvious thing – and also aesthetic comfort, such as a sense of spaciousness and light. While delight, which is a delightful word, is relatively self-explanatory.

I don't want to project manage and produce sequencing schedules and spreadsheets. The only critical path I want to see is the one leading to my front door.

I find that I've come to like this definition less and less, partly because it's a definition that needs some kind of translation out of the language of the architect and partly because it's not a very adventurous or demanding definition.

Unlike definition number **four**, which uses, very neatly, four ideas. Over the past seven years of *Grand Designs* I've been slowly simmering away a sort of checklist, a list of basic requirements that our homes ought to provide. And after quite a lot of evaporation and reduction, I've condensed it down to four basic principles of good domestic design, which, other than providing shelter and being well built, our houses should do really well. I'm a little nervous about sharing this world view with you because I'm worried that I've left out a fifth crucial principle that I haven't even thought of. But if it hasn't surfaced in nearly eight years, it never will, so here goes:

Four basic principles of good domestic design

First, our architecture needs to be contextual (this is really the 'place' argument of the 'people and place' definition above). We need to be building houses that look as though they belong uniquely to one place, in styles and materials that reflect their region and location and their setting. Like cheese. Britain now has more local cheeses than France. And like cheese, I don't mean that houses need to be pastiches of local vernacular building types, like Cathedral City Cheddar, but that their design ought to pay respect and homage to where they are, the history of the place, its landscape and culture. Sounds difficult perhaps, but that's why architects train for seven years: employ a gifted architect and this first principle of making houses look as though they belong becomes very easy.

Second, they need to be sustainable. This is a big empty word, these days, but one that I think can be basically defined in terms of how we responsibly use the planet's resources. So that means houses that have a minimal environmental impact in construction and a minimal impact in use: with high levels of insulation, minimal heat loss and clever energy and resource technologies such as heat-recovery systems, greywater recycling and district heating systems powered by alternatives to fossil fuels, such as biomass. All this, of course, means that the current building regulations are way below what are really necessary. As to the government's peculiar term 'sustainable community' that incorporates social and economic sustainability, that's a concept that goes way beyond housing design into the realms of town planning and urban design. But I'll tell you one thing: unless you can get the first principle of contextuality right (see above), you'll never get people feeling that where they live has any real sense of place or localness. A community won't 'sustain' unless it has some sense of belonging.

Third, our houses need to be contemporary. I don't mean that we should all live in glass boxes (see ideas 1 and 2). In fact, our architecture should be as diverse and mixed as possible. Nor do I mean a list of all that macho category-five-future-proofing-cable gizmo stuff, like mood dimmer systems and remote-controlled electric window blinds connected to your own satellite weather station. People think of that as being so very up to the minute, but of course it'll all go wrong in a year's time and date very quickly. No, by contemporary I mean that it should

facing page: **Buildings are for people to happily inhabit: Rob and Nicki Parkyn's Cornish house.**

serve our contemporary needs and be flexible enough to accommodate small changes in the way we live. Which sounds a bit abstract but in practice it means, right now for example, large kitchen-diner-living rooms that have TVs, sofas and family socializing spaces all in one, with quiet rooms for reading and homework, or large storage rooms (AKA the garage) for surfboards, bikes, hobby stuff or the dog. And bedrooms large enough to get a good-sized double bed in as well as a pair of human beings. As for change in the future, a design might incorporate multiuse space or walls that can be removed or repositioned easily. And for the gizmo junkies, trunking or cable conduits that are easy to rewire with whatever the future has to bring.

Fourth (and I have to thank the designer Tom Dixon for enunciating this point so succinctly), our homes should be bloody nice places to live. They should make us feel like better, healthier, kinder, more civilized human beings (which is really my first definition again). They should detox us of the ravages of the world. They should give us space and light and generous headroom. There should be plenty of storage for life's clutter and an ease and joy in interacting with the building that continues day after day. In short, we should have a positive, enriching relationship with our home.

Future perfect

These four principles are actually not much to ask for. They are certainly not much to ask of any architect, who would consider them as something of a basic brief when designing a house. Nor, indeed, are any of the above four definitions (pages 20–1). Yet, very often, in the pursuit of a Balinese wet room or an American walk-in fridge, it's these gentler, more philosophical ideas that clients forget.

I described O'Donnell and Tuomey's attitude that they bring as a practice to their work, but I should also mention someone else I know, Karen Helsingen, who built a £250,000 house with her husband in the conventional relationship with a builder and architect. This was to be their new family home in Herefordshire and represented something of a sea change for them, having moved from a city to the countryside. What struck me about her brief to her architect was how it wasn't bound up with the usual lists of features and white goods, but instead was really a list of experiences that she wanted to get out of the place. Things like 'a view of the sunset' and 'being able to step outside and pick flowers' but also a few more mundane requirements like 'getting out of bed and not having cold feet'. If it were me, I might have added a long list of more practical requirements, like self-cleaning children, but I must resist going on, because her list was really much more abstract. As I say, a set of experiences rather than physical features. Particularly intriguing was her explanation of the brief: not as a 'wish list' of things she's always dreamed of having but as something more reliable; a list of things that, through reflection and assessment, she'd come to realize had made her happy in the past.

Now that's more like it: someone who dwells a little more on the past than on the never-neverland of the future. There is an awful tendency among self-builders and anyone commissioning a building to go overboard. After all, if you're borrowing 18 times your income, why not borrow 20? Hey! And then go over budget to boot, and borrow a bit more? The whole process seems to me to be built around the one principle of putting your life in hock to the future. The mortgage will need paying somehow, sometime, but don't worry because the new house will somehow, sometime springboard you into a brighter, happier, more affluent future (though I've never seen any statistics to support this blind optimism). The design will change the way you live and is bound to make you tidier, neater human beings (some hope). You could also wreck the immediate future by building it yourselves and putting your social and private lives on hold for four years while you move into a caravan with the kids. It'll be fun! And to top it all, why not irreversibly embody in the design and layout all the details, gizmos and features that you've never had but dream of owning? Even if you have no real idea whether they'll make you a happier person who's more at ease with the world. Now there's risk.

> There is an awful tendency among self-builders and anyone commissioning a building to go overboard.

The worst case result for this kind of approach is a house that you can't afford to live in or heat properly; that prevents you, through penury, from going on holiday for five years; that takes for ever to finish; that is alienating; and that

begins to date from the moment it's finished. I really can't think of a more depressing recipe for disaster.

Meanwhile, Karen got it right in my view. She spurned a long list of expensive appliances and features and so scaled back her budget. Her 'dream' was founded in what the finished building would give her, the relationship it would create with her, and not in the 'experience' of building it herself or of stuffing it full of modern goodies. Her home was relatively simple when finished but her architect had delivered those experiences she wanted. Non-material experiences, too, like warm feet in the morning. Result: happiness. Which is all any of us really want in the end, isn't it?

The danger of adopting such a retrospective and modest approach is that there's no ambition in it. It's important that, just as your architect must deliver your dream fully formed and immeasurably amplified, you, too, have a responsibility to inspire them and not just provide them with a shopping list of what you want. Another client I recently met, Nicholas, commissioned a highly contemporary eco house in oak with a circling cob wall, for which the brief was a simple three-line Haiku poem that he wrote, using words like 'ecological', 'light', 'tranquil', 'engaging', and so on. Admittedly, his architect was initially baffled and quite keen on amplifying such a minimal brief. But the point here is that the poem represented a sort of lyrical leitmotif that could continue to run through the brief as it developed. It was a piece of poetical philosophy that could lift the day-to-day requirements that the client wanted out of the mundane.

The hypnotic allure of the new

This leads me into a slightly thorny area that we will have to deal with because it's all about what you want out of this home and, for that matter, out of your life. It's something I've alluded to and skirted around and poked at and it's this troublesome walk-in fridge/wet room/let's-buy-a-new-Audi-while-we're-about-remodelling-our-life thing. There is no avoiding the fact that a new house is a new object, a new material possession. It's also (probably) going to be full of more new possessions. It's going to be a Predictable Shiny New Thing (or PRESNT for short).

facing page: **Butterwell Farm in Cornwall by Charles Barclay.**

We reward or treat ourselves with new PRESNTs all the time. Men of a certain age buy clunky great expensive Swiss watches that look as if they've been assembled from bits of stainless-steel hand grenade. Most women I know enjoy being given jewellery. My daughter, in particular, loves glittery pink Barbie dressing-up kits. New cars, new DVD players, new boats, new shoes, new conservatories – they are all baubles: shiny, alluring objects to own that glimmer with newness. As far as I can make out, there is not a human being on the planet who is immune to the charms of PRESNTs. We seem to have a genetically coded weakness for new things.

New buildings are some of the most expensive PRESNTs going. National governments, companies and institutions are all as vulnerable and open to seduction as the rest of us when it comes to the glamorous appeal of the new.

> If you place all the value in your design on technology or on the material possessions inside your home, the sooner it will start to age and feel out of date.

And yet what you have to remember is that the moment something comes into existence, it starts to age. Just as cars tarnish and rust, buildings grow algae and moss, and skirting boards get scuffed (that is, after all, why they were invented). I can begin to see the allure of women's jewellery here: diamonds and gold and sapphires and emeralds don't decay or rust. That's where their value lies. Whereas, if you place all the value in your design on technology or on the material possessions inside your home, the sooner it will start to age and feel out of date.

For that matter, we're not just talking about how things look and feel. The bitter experience of living with five children and frequently having to dispose of/repair old failing iPods, hairdryers, DVD players and Gameboys tells me that the more technologically complex your life is and the more toys your house has (alarm systems, CCTV, remote heating controls, automatic window closers, lighting mood systems), the more likely it is to all go wrong or break down. Even walk-in fridges fail – and often sooner than you'd think. So my advice is to try to keep life simple wherever possible and really do ask yourself whether that scene-setting computer program is necessary or whether you couldn't manage just as well with a light switch.

facing page: **Castle Gwydir in Wales.**

In fact, I feel like writing a manifesto in defence of the light switch and of older acquired furniture and, for that matter, of buildings that age gracefully and change character as they do so. I've never been a fan of those homes where the owners have sold all their possessions only to buy a whole new raft of 'contemporary' objects and furniture. The ageing process starts immediately, so much so that two years down the line the home looks staged, like a vintage set, and not wholly real. But, most importantly, the most interesting homes are those where you can sense the passage of time; where the clapboards are greyed and mossy; where the interiors don't look like magazine spreads but where objects and furniture have been collected over time. What turns a house into a home is our use of the place and the idiosyncratic mark we make on it with our mess and our cherished belongings, and the added resonance of who we are and where we've been. The best homes are not style statements, but autobiographies.

I suppose that I'm making a plea here: for less of the headlong, headless pursuit of the new at the expense of the established; for less material ownership and instead more experience; for 'home' as a place that can be incorporated into our lives and our personal histories, as opposed to 'house' as some launch pad for a new glossy future. It's been said that we live in an age where we are no longer judged by who we are but instead by what we have. I think that your home should clearly be a statement of who you are.

How to be a perfect client

Admittedly, in the eyes of any builder or architect there may be no such thing as the perfect client, so there's still a chance for you to prove them all wrong. To help you I've listed a few guidelines here. They're not arbitrary. Like most of the advice offered in this book, they're based on years of experience watching a lot of people get it wrong – and a few who get it right. So:

DON'T employ the first architect you meet (see page 47)

DO interview as many as you think you need to and look for someone whose work you like and admire and whose view of the world most closely fits your own.

DON'T try to do everything yourself.

DO employ as many professionals as possible. They know what they're doing. (So much so that there's an entire chapter devoted to this point coming up; see Everybody Else, page 44).

DON'T treat your architect like just some other service provider or contractor.

DO make friends with them. This is the person who is going to deliver not just your home but perhaps 20 years of your future life. You have to know and trust each other for that to be a success.

DON'T think you can do the job without a good architect. They're trained to take your brief and deliver it back to you as a fleshed-out, living, breathing building.

DO employ your architect on a full-service contract. They'll hold your hand through the whole process and give you a properly finished home with the minimum of compromise.

DON'T commission a house as if you were buying a new car. It's not a transaction but a process. And don't think of it as a giant exercise in interior design.

DO remember that you're responsible for bringing an original piece of bespoke design into the world that isn't just for you but for the house's future owners, your neighbours, and anybody who has to look at the building or will have an encounter with it. It's a very creative act of patronage on your part but it comes with some social responsibilities.

overleaf: **Future Systems' earth-sheltered 222 House on the west coast of Wales.**

How to put a brief together

You will notice I haven't used the word 'write' in this title, because a 'brief' can take all kinds of forms. It can be Nicholas's Haiku poem, or a set of photographs, or a series of conversations. But here are some more advisory dos and don'ts.

DON'T put together a visual brief entirely composed of tear sheets from *Homes & Gardens* (or *Grand Designs* magazine, for that matter) showing interiors and fitted kitchens that you like.

DO put together a visual brief with photographs of family, previous houses, pets, holidays and anything that has ever inspired you – as well as the odd tear sheet.

DO though, subscribe to some home design magazines. It's a good idea to take two or three on subscription. Most of them, of course, illustrate sickeningly tidy new homes, as well as the odd alteration or extension. Perhaps most importantly, they'll provide lots of budget and technical information and useful contacts.

DON'T rely entirely on home interest magazines for your inspiration.

DO visit the RIBA bookshop in Portland Place in London and scour inspirational architectural books. For technical inspiration, visit manufacturers or go to exhibitions. Most of the major manufacturers of building materials, design products and services exhibit at these shows and they're all very anxious to sell their products and services to you direct. The Building Centre in London also has a shop and regular exhibitions.

DON'T treat your meetings with your architect like business appointments with objectives and deadlines.

DO get to know them, go on walks together and even visit buildings together. For that matter, you should be taking a lively interest in the design of other people's houses. Visit friends, visit show houses; make notes of design features, colours, finishes; ask friends how they get on with a particular layout. Compare their requirements with yours and ask yourself whether their design solutions will provide an answer for you. You'll find that most people are only too happy to talk about how they've arrived at a particular layout or a design for their new home.

DON'T try to remember everything you see and learn. That's impossible.

DO write everything down: comments, friends' opinions, and technical information. And take photographs – lots of them. I suggest that you have a scrapbook for keeping photographs, ideas, articles and snippets of information. In a way, this less-than-beautiful assemblage can often form the perfect brief.

DON'T deliver a neat written brief and some pictures and think that's it. A design develops over time in response to input.

DO be prepared to have lots of meetings to discuss and change things. Remember the couple who saw nothing but flow diagrams of how they use their home, in a design process that lasted six months. So be prepared to engage.

DON'T leave the layout and plan entirely to your architect.

DO measure up the furniture that you want to relocate. Measure the plan footprint and height of furniture and fittings required and think about space around furniture for access.

Needs

Given that you'll be researching ideas for your brief right at the beginning of your journey, often before employing an architect and perhaps even before finalizing a site, it's important at this stage to keep all your ideas flexible. But recording your actual requirements is also important. For example, if you're going to have a dining room, you've then got to decide how many people it will seat: maybe a family of six with the option of going to ten. Or maybe you're curmudgeonly and only need it to seat two at any given time. So try to establish, for each room, a range of your needs (comparisons with your existing home can be helpful). This means somehow projecting yourself ten, twenty or even thirty years into the future and predicting what life will be like then. It may be hard to do, but you may have to account for:

1 Having children. Or more children. Or grandchildren foisted on you.

2 Retirement. The need to consider access and mobility (a bedroom on a ground floor, for example, or a shower room) and whether, once you're at home with your partner, you will each need some private space far, far away from each other.

3 Business. The possibility of running your own business from home or even providing a home office for your company.

4 Annexe accommodation. Providing accommodation for an elderly parent, a nanny or a teenager (if necessary, I would put all three in an annexe).

5 Children fleeing the nest. This might mean retaining your home but using and heating it in such a way that you can occupy a portion of the building comfortably, shutting parts of it down, say, during the week.

6 Divorce. Many people find the trauma of selling up just too much on top of the additional stresses of marital break-up. One novel solution is to design the house so that it can be split into two. Extreme, I know, and a little like writing a prenuptial agreement, but it has worked for some.

facing page: **Make provision for a stair-lift and see what happens when the grandchildren come to visit. Here, in Alex Michaelis's west London house, the kids can take the stairs – or the slide.**

The three ages of parenthood

In an attempt to cover all the remaining ground and also to prompt you as you devise your brief, here is a list based on Shakespeare's Seven Ages of Man, which I've managed to condense down to the Three Ages of Parenthood. The conclusion I've come to is that space is vital right the way through the whole experience of bringing up children, and flexible space at that: a nursery might transmogrify into a playroom, into a TV room, into (perhaps, one day) your study. If you get things right at the start, you will find that your home can expand, contract and adapt to the changing needs of you and your children. And with any joy you'll even be left with somewhere you can still enjoy living in after they've all left home.

1: THE EARLY YEARS

Space is always a serious issue with children. You need to think about storage of equipment: nappies, push chairs, play mats, toys, trains, doll's houses, Barbie dolls, balls, boxes of bricks, bikes, skateboards, outdoor play equipment. Any parent will tell you that the list is endless and many of the objects vast, ugly and cumbersome. Then there's the need for a convenient downstairs WC, ideally, equipped with a bath or shower. And a separate playroom, somewhere that can be easily supervised from the heart of the home where you'll spend most time, be that the living room or the kitchen. If your family, like so many, is a second one, bear in mind that the needs and interests of five-year-olds are not those of a teenager. You might have to consider giving up your study, library or gym for a few years and converting it into a 'teenage playroom' until the older children have moved on (see my note about annexes above).

2: THE MIDDLE YEARS

'We had the dream of family walks together, but all the children wanted to do was to have their friends over and play on the computer!' When you have children, where you live is important. If you're out in the sticks in the middle of some idyllic countryside setting, you're inevitably going to spend a lot of time toing and froing collecting and delivering your children and their friends. I can personally vouch for this, since at weekends I have a second job as an unpaid minicab driver.

Of course, this all becomes less of an issue as children get older and can drive, but by then they're considering moving away anyway.

And don't just think about where you want to live. Think about where you want to educate your children. The length of school runs can make or break family happiness; in my view, these shouldn't be more than 20–25 minutes in the car. Even a 10-minute drive to school is three hours 20 minutes per week.

3: THE LATER YEARS

The shocking truth about children is that they grow into adults – huge lumbering creatures that loom through your house, blocking doorways and requiring a sofa each, all day every day, for weeks. To keep them from spending their entire lives in bed, you'll need the space for giant TVs or, heaven help us, a home cinema; personal computers (probably plural); PlayStations; video, CD and DVD storage; mini-disc player and so on. And you'll need some kind of utility room for them to keep their coats, hats, outsize shoes and boots. Not unreasonably, teenagers want to have space of their own – reasonably generous and private bedrooms and, particularly, places to study.

Think also about bedrooms with en suite facilities that can take the pressure off one bathroom. And if you have just one bathroom, make it as large as possible and incorporate a separate shower. Remember, the older children get, the more time they'll spend in the bathroom, so at the very least install a separate WC in the house – with a powerful fan.

> The shocking truth about children is that they grow into adults – huge lumbering creatures that loom through your house, blocking doorways and requiring a sofa each.

Handy helpful list of tips

Finally, not so much an Anorak's checklist, more of a High-Vis-Vest checklist for your brief of almost every feature, accessory and technical component you could hope to include:

Kitchen

Aga/range cooker;
breakfast bar;
views of garden;
worktop;
storage;
raised-level dishwasher
   and oven (very useful);
built-in scales;
induction hob;
water filter;
adjacent conservatory;
separate breakfast room/area;
central island;
integrated recycling bins;
American-style fridge-freezer;

farmhouse-style kitchen
   (with a table at the centre);
'children's' food cupboards and
   fridge for milk, juice, cereal and
   bread near to the dining table;
sink preformed in the worktop;
hideaway kitchen (within
   cupboards);
central hub kitchen;
   (overall design layout);
kitchen within a kitchen (magic
   triangle of sink/fridge/cooker
   accessible only to cook;)

music and television;
computer for Internet shopping;
pull-out pan drawers;
task lighting;
view of adjacent playroom
   and of children's outdoor
   play area;
easy access from street/
   drive for shopping;
flexible fenestration like sliding
   doors out to a terrace;
sofa and easy chairs
   for open-plan
   living/dining/kitchen.

Utility room

Dryer/washing machine
   with feeds and outlets;
sink;
clothes cupboards;
clothes horse;
coat hooks;
boot storage or even separate
   boot and coat room;
possibility of first-floor utility
   room adjacent to bedrooms;
thick door to prevent noise
   penetrating the whole house;
built-in dehumidifier;
dog washing/hosing down;
shower (flush to floor);
rubber flooring;
more cupboards;
and yet more cupboards.

Sitting room

Island fireplace (double aspect);
2-amp lighting circuits for table
 lamps;
home cinema, or no home cinema
 or television to form a quiet zone
 in the house;
built-in hi-fi;
inglenook fireplace;

flexible fenestration like sliding
 doors out to a terrace;
blinds enclosed in glazing units;
power points in the floor;
chimney/fireplace as room divider;
changes of level;
task as well as mood lighting;
curtains, blinds or shutters;
bookcases;
desk and hidden home-office
 provision.

Lobbies

Feature staircase;
balustrade;
galleried landing;
guest WC;
guest coat storage;
master switches by the front door;
security system;
large doormat;
nearby bike storage;

entrance porch (to create an airlock
 and some privacy);
sofas or built-in seating for guests;
wide halls to accommodate
 both incoming and outgoing
 traffic.

Bedrooms

Walk-in wardrobes or separate
 dressing area;
walk 'onto' balustrade;
en suite dressing room;
en suite bathroom;
music;
dimmed lighting;
sleeping platform
 (at mezzanine level);
bedroom fireplace;
bathtub in the bedroom;
central bed headboard
 with underfloor wiring;
cupboards;
blackout blinds/curtains;
view from the bed;

view of the stars through
 a skylight;
Wilton carpet underfoot;
privacy;
reduced noise, acoustic insulation
 in walls, floors and ceilings.

Children's bedrooms

Feature bed/bunk bed;
study facilities;
en suite facilities;
storage and cupboard space;
interconnection with other
 bedrooms;
blackout blinds/curtains.

Bathrooms

Twin basins;
wet room tiling;
rubber floor and walls;
'flush to floor' shower;
shower with multiple body jets;
double-ended bath;
Jacuzzi;
book and magazine storage;
children's toy storage;
sunken or raised bath;
view from bath or WC;
shared en suite (between
 multiple bedrooms);
extractor fan;

cupboards;

Wilton carpet underfoot
(I recommend this when
stepping out of a bath or
shower);

mood lighting;

make-up mirror with all-round
lighting;

TV or music;

easy chair.

Communal/corridor spaces

Sun pipes;

galleried landing;

linen cupboard;

clothes storage;

book storage;

extra-width landing for desk
or furniture.

Heating/climate control

Underfloor heating;

trench heating;

underfloor heating in bathroom;

programmable taps;

pressurized domestic hot water;

weather compensation/
optimization system;

air conditioning/comfort cooling;

condensing boiler;

heat exchanger;

groundsource heat pump;

deep underground
geothermal heating;

solar panels for hot water;

photovoltaic panels for electricity;

in-line water filter;

in-line water softener;

flat, invisible or feature radiators;

zone controls;

thermostatic radiator valves;

vent stacks to roof;

outside pond next to the
south elevation for maximum
'stack effect' passive cooling
in summer.

IT

Phones;

home network;

TV/satellite signal distribution;

home automation;

home entertainment;

auto-lock entry phone.

Solar/energy conservation

Rainwater harvesting;

greywater recycling;

reedbed sewage system;

composting toilet;

photovoltaic roofs;

solar hot-water heating;

concealed fluorescent lighting;

low-emissive glazing;

triple glazing;

windmill;

super-insulation;

heat exchanger;

groundsource heat pump;

district heating (for multiple
dwellings);

concrete or limecrete plinth
for maximum thermal mass,
combined with south-facing
glazed elevation;

sun room or conservatory
as additional 'glazing unit'
to the south;

'brise-soleil' or sunshade;

deciduous tree to the south
of the building to shade in
the summer only;

overhanging roof.

Security

Alarm systems;

home security systems;

CCTV;

exterior lighting;

dog.

Outside

Outdoor hot tub and/or shower;
outside water taps;
auto-irrigation systems from
    greywater and rainwater storage;
underground water storage;
electric gates;
swimming pools (natural or other);
decks;
terraces;
garden;
veranda;
roof garden;
drive;
parking;
lighting;
landscaping;
exterior maintenance;
access;
security;
shed;
waste management;
recycling;
sewage;
runoff of groundwater.

During construction

Lorry access;
site security;
site office;
storage area for deliveries.

Miscellaneous items

Underfloor safe;
basement;
cellar, wine cellar;
central vacuum system;
gym;
games room;
study;
first-floor or roof-top conservatory;
exposed structure ceilings;
deep window reveals or
    splayed window reveals;
internal window shutters;
sun pipes;
self-coloured finishes such as
    plaster, concrete and wood;
concealed doors;
full-height doors or openings;
suspended or 'floating' walls;
structural glass;

roof lights;
concept of upside-down living,
    with the living area on
    the first floor;
secret room;
stable;
garaging;
bothy or ancillary stores;
annexe;

balconies;
'honesty of construction'
    build methods;
lift for people or small one
    just for food;
first-floor kitchenette;
first-floor utility room;
provision for pets.

The most important piece of architecture that you are ever likely to experience is your own home; it's the one piece of the built environment in which you will spend most of your life. Architects spend seven years training, while all the time considering the creative use of space, the appropriateness of materials, the importance of detail and the intangible improvements in human wellbeing that occur in a well-designed environment. So we ought to use them.

I've seen plenty of projects not designed by architects and they all share one thing: a horrible clumsiness. At best, houses that are not professionally designed are diKcult to negotiate and feel stunted or poorly detailed. At worst, they feel like shopping malls or airports: buildings in which the mechanical functions of the place, like air-conditioning ducts, seem to take precedence over human enjoyment of the place.

And, of course, a good architect is free. They'll cost you no money in the long term, because if they're any cop they'll take your vision of what you want, strain it through their own peculiar and labyrinthine mind, prod and knead it gently with a pencil and rubber, and then bake it at 200°C in their CAD (computer-aided design) software before delivering it back to you in a pretty box. You don't get the raw ingredients of a building with a good architect, you get the perfumed, heavenly experience of the finished syllabub: you get your vision fully wrought

facing page: **Springett House, London, designed by Matthew Springett and Kirsteen Mackay.**

and resolved and handed back to you perfectly formed and larger than you ever thought possible. As a result, your home will be better and therefore worth more, and you will be happier. Now that is a secret of life.

One of the secrets of a happy build is, perhaps, to rely on as many chosen professionals as you can possibly afford. In my dream build (location = where I live now; budget and schedule = on the nail; likelihood of ever happening = zero), I'd employ an architect on a full-service contract to hold my hand all the way and produce every detail drawing necessary. And I'd like a good working relationship with the structural engineer. I think I'd also like a quantity surveyor involved to cost the project – who might not be the same person as the project manager, whom I suspect I might want to employ separately from the builder or subcontractors – for the reason that a house is such a complicated machine, involving so many different trades, that it requires the supreme coordination of one human being. If my planning application ran into trouble, I might need a planning consultant, and if my plot turned out to be on the site of a medieval dung pit, I might want to employ a soil engineer, if only to dig it out.

You don't get the raw ingredients of a building with a good architect, you get the perfumed, heavenly experience of the finished syllabub.

But why employ so many people, each with their own speciality? For the same reason that I'd employ a builder and even a labourer on the job. I don't want to design my own house, thanks very much. I don't want to have to produce the reams of drawings required (I've done it once for a conversion and extension and that was enough). I especially don't want to be a labourer on my own house. I don't want the hassle of coordinating plumbers and electricians and plasterers, most of whom will be complete strangers whom I'm never likely to meet again. I don't want to project manage and produce sequencing schedules and spreadsheets. The only critical path I want to see is the one leading to my front door.

I want to be a good client. I want to have the luxury of being able to pore over drawings and consider and reflect. Any one-off building – and particularly a private house – will evolve as a design through the build, and it's important as a client to find the time to stand back and be part of that design process.

It's important, too, to find the time to make the thousands of small decisions that are needed by the architect and builder, about detail, finish, specification and layout. I can't overemphazise how many decisions there are to make; just being a client can be a full-time job.

> It's important as a client to find the time to stand back and be part of that design process.

Of course, you may choose to ignore this advice. You may feel impelled to do some of the construction work on your own house. You may just not be able to afford to do otherwise. In which case you need to remember the fallback position. The very best buildings result from collaboration between three key committed people: the client, the architect and the builder. Try to reduce it to less than three and you face the real danger of failure.

Finding an architect

There are good and bad architects; and it's worth spending a lot of time and even a bit of money to make sure that you have found the best. You may have to do a lot of foot slogging, visiting different architects' offices to find a design approach that really appeals to you. As with any professional, there is a tendency for an architect to get a name for a particular type or style of building: eco, vernacular or minimalist, for example. But the really good architect is the one who's able to rise to all the challenges set out in your brief that I discussed in the previous chapter (see pages 34–43), and who's able to fashion your hazy, wobbly dream into a perfect reality.

WHERE TO FIND YOUR ARCHITECT

1 The Royal Institute of British Architects (RIBA) is based in London and has a whole selection of services and useful information available for both architects and clients. It runs a client's advisory service (CAS) that will match a client's requirements to an architect. The RIBA also has regional offices all around the country, which will have closer contact with individual members or practices in your area. It is worth finding out from the RIBA where your nearest regional office is, then calling them to ask for advice as to architects or architectural practices that might be able to help.

2 The Architect's Registration Board (ARB) is a register of all qualified architects. In order to qualify as an architect and use the word architect after their name, he or she must complete extensive training and clock up appropriate experience. In order to be a member of the RIBA, an architect must have qualified and therefore will always be on the list of registered architects.

Note that there are a few architects who are registered and entitled to use the word architect after their name, but have decided not to join the RIBA and therefore do not use the initials RIBA. The majority of architects, however, are members of the RIBA, as they want the umbrella of the services and support that the RIBA offer. Note, also, that building designers or architectural technicians are not qualified architects.

3 Word of mouth/recommendation is probably the most common form of reference for architects. A good architect will build a reputation by doing jobs well, and as a potential client you can't beat talking to two or three former clients who have been through the process.

4 Magazines are also useful sources of information about architects. Make notes of homes that you like, find out who the architect is and contact them.

5 Most architects or architectural practices will have a website where they'll have expounded their design philosophy and where you can see examples of their work. These online portfolios can save enormous amounts of your time and they really are worth consulting in your first round of research before you make contact. The site may also have a published sustainability statement.

Remember that most local architects will make an initial visit at no charge in order to assess the site and what it is that you want. But it's always worth asking, before you arrange an appointment, on what basis fees will or will not be payable. It will also give some comfort to the architect that you are not expecting something for nothing.

facing page: **The mixed-tenure Bloembollenhof estate in Vijfhuizen, Holland, by S333, where great architecture and clever social planning go hand in hand.**

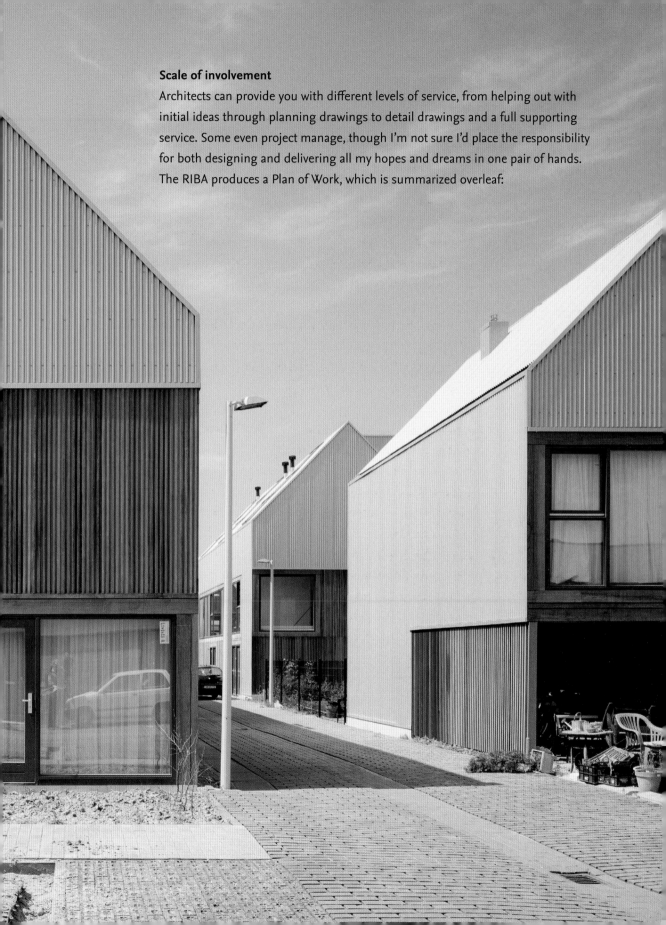

Scale of involvement

Architects can provide you with different levels of service, from helping out with initial ideas through planning drawings to detail drawings and a full supporting service. Some even project manage, though I'm not sure I'd place the responsibility for both designing and delivering all my hopes and dreams in one pair of hands. The RIBA produces a Plan of Work, which is summarized overleaf:

The RIBA Plan of Work divides the design and construction process into convenient Work Stages. Architect's services and fees are usually based on these work stages.

Feasibility

A APPRAISAL

Identification of the Client's requirements and of possible constraints on development.
Preparation of studies to enable the Client to decide whether to proceed and to select the probable procurement method.

B STRATEGIC BRIEF

Preparation of Strategic Brief by or on behalf of the Client confirming key requirements and constraints.
Identification of procedures, organizational structure and range of Consultants and others to be engaged for the Project.

Pre-Construction

C OUTLINE PROPOSALS

Commencement of development of Strategic Brief into full Project Brief.
Preparation of outline proposals and estimate of cost.
Review of procurement route.

D DETAILED PROPOSALS

Completion of development of the Project Brief.
Preparation of detailed proposals.
Application for detailed planning approval.

E FINAL PROPOSALS

Preparation of final proposals for the Project sufficient for coordination of all components and elements of the Project.

F PRODUCTION INFORMATION

F1 Preparation of production information in sufficient detail to enable a tender or tenders to be obtained.

F2 Preparation of further production information required under the building contract.

G TENDER DOCUMENTATION

Preparation and collation of tender documentation in sufficient detail to enable a tender or tenders to be obtained for the construction of the Project.

H TENDER ACTION

Identification and evaluation of potential contractors and/or specialists for the construction of the Project.

Obtaining and appraising tenders and submission of recommendations to the Client.

Construction

J MOBILIZATION

Letting the building contract, appointing the Contractor.

Issuing of production information to the Contractor.

Arranging site handover to the Contractor.

K TO PRACTICAL COMPLETION

Administration of the building contract up to and including practical completion.

Provision to the Contractor of further information as and when reasonably required.

L AFTER PRACTICAL COMPLETION

Administration of the building contract after practical completion.

Making final inspections and settling the final account.

(Source: RIBA)

The Plan of Work isn't comprehensive but it summarizes the main areas where an architect gets involved on a building project, and it's a very good basis for negotiation. In practice, if you employ an architect on a Full-Service Contract (the whole deal, which is the one I would always opt for) you can expect them to keep a really beady eye on construction and carry out periodic site inspections. This isn't site supervision, though – that's the job of your project manager; in order to supervise a project, an architect would have to be on site all the time. On average, expect them to visit the site once a week, depending on the stage of construction. They carry out these inspections to make sure that the works are generally in accordance with their written and drawn specification and to redesign or modify the design with you as work progresses. Remember that every original new piece of architecture is its own prototype and that things will go wrong and bits of the design may need reworking. Also remember that any original design needs input in the building stage: there are some decisions that just can't be made on paper and which you need to make on site with your architect.

Throughout the project, your architect will have been signing certificates authorizing payment at various stages and will produce a final certificate at the end of the job. (This is often a requirement of the person providing funding or the mortgage company.) Provided that all these certificates are signed off, you have, in effect a form of building guarantee – if problems occur at a later stage, the architect's professional indemnity insurance (see later) will cover you against any design defect. It won't, however, necessarily cover any construction defect.

BUILDING INSURANCE

It's an extremely sensible idea to have some building insurance. This covers you against most (you need to read the small print) building defects, usually over a period of ten years from completion. It might appear expensive but my advice is to get it, remembering that you need it in place before you start. Some form of building insurance or guarantee is something that, if you come to sell the place, more and more people expect to see when buying a home less than ten years old.

> Remember that every new piece of architecture is its own prototype and that things will go wrong and bits of the design may need reworking.

Are architect's fees worth it?

Well, this question will be still being debated at the Last Judgement. It all depends on whether your architect has taken the loose bag of ideas that you presented him or her with and has in turn presented you with a fully detailed, fully wrought thing of total, mind-blowing perfection. If so, then, um, yes. In my view, architects are usually worth every penny. The general perception, of course, is that they make a lot of money. As I write, I'm desperately trying to persuade/cajole/steer my 17-year-old son into becoming a hedge fund banker, because he's got it into his head that architecture is the thing he wants to read at university. What I suspect is that if he qualifies and goes into practice, he won't, like nearly every architect I know, design a whole building of his own until his late thirties. What I know is that for the foreseeable future I'm going to have to support him, not the other way around, as I was hoping.

Work it out. Your architect will spend an enormous amount of time on your project; they'll produce reams of drawings; they could be working with you for one to two years. At an hourly rate, their contribution will end up way below anything you're likely to be paying the other professionals on the job. More than likely, it'll less than what your plumber is paid.

The initial design input of an architect is probably the most valuable input. So many of the home builders I meet decide that they will use an architect simply to get planning permission and then they'll get a technician or even the builder to work out construction details and manage the project on site. Now that's a pity. Because it is so important that a design is allowed to mature and develop over the course of a project – even in construction, as I've said – and paramount that the integrity of the design concept is maintained, right through production of working drawings to the detailing and selection of materials. Not to mention the attention to the thousands upon thousands of tiny details that collectively can turn a new house into either a RIBA medal winner, or a dog.

So, if you are going to take the plunge and employ your architect right through to the sweet end, I, for one, applaud you. You can get some advice

In my view, architects are usually worth every penny.

on fees from the Royal Institute of British Architects; it produces a useful document called 'Guidance to Clients on Fees', which you can be order (see the RIBA website: www.riba.org). These days the RIBA, as with most professional organizations, is reluctant to give specific advice on fees, but it will provide a range of typical fees that will give you a rough guide. It's up to you, the client and the architect, to negotiate an exact amount. You'll find that for a new home project with a total construction cost of around £300,000, architect's fees will typically work out between 10 and 15 per cent of the total construction cost, for a full involvement (RIBA Plan of Work, stages C–L).

Other professionals

QUANTITY SURVEYOR

A quantity surveyor will be able to advise you on building cost at the beginning of the project and will also help to monitor the costs throughout the course of the project, keeping tabs on any extra costs as well as any savings that are made along the way. Most new home design projects should have quantity surveyor involvement at some stage – the degree to which they are involved will depend on the complexity of your design. Most are happy to work on an hourly basis, though fees are usually based on a percentage of construction cost. Find one through the Royal Institute of Chartered Surveyors (RICS): www.rics.org.

STRUCTURAL ENGINEER

Although your architect likes to take the credit for 100 per cent of the brilliant gravity-defying and ergonomic features of your new home, the chances are that the design will only have 1) met building regulations and 2) been technically buildable thanks to the hard slog put in by your structural engineer. Structural engineers are the unsung heroes of modern architecture, their presence quietly demanded, thank goodness, by most local authorities. Their charges are based on an hourly rate or can be calculated as a percentage of the construction cost. Contact the Institution of Structural Engineers (IStructE): www.istructe.org.uk. You may find that your architect has a good working relationship with just one or two engineers in particular.

facing page: **You should even use an architect to design an extension: this is the interior of Henning Stummel's radical three-storey bathroom extension to a terraced house in London. It won him a RIBA medal.**

SOIL ENGINEER

If you're building on a 'virgin' site, soil investigation is imperative. A structural engineer will usually involve a soil specialist, who will check and analyse soil taken from trial holes on the site. Soil tests cost a few hundred pounds, but they're always worth carrying out at an early stage to establish any unusual site conditions that could affect construction costs: you may get away with some shallow trench footings; you may in the end have to drive 80 piles down 35 metres, each at a cost of £29,000. Who knows? Only a soil engineer does.

PLANNING CONSULTANT

Don't underestimate how valuable a good planning consultant can be, not only in helping you get permission for your home in the first place, but also in helping with difficult planning conditions. Advice is usually given on an hourly basis and guidance on fees can be obtained from the Royal Town Planning Institute (RTPI). Visit its website: www.rtpi.org.uk.

PARTY WALL SURVEYOR

No man is an island, and few properties are. Almost every plot abuts another one. And if your site is in a built-up area, the chances are that your boundary will be near other buildings. Now, if you lived in France, Napoleonic law would provide a neat solution in writing for every conceivable party wall circumstance. In the UK we prefer to operate our legal system by negotiation and precedence, hence the unique speciality of the party wall surveyor. If you're twiddling your thumbs on a rainy afternoon, you could entertain yourself by reading the Party Wall Act on the Department for Communities and Local Government website: www.communities.gov.uk.

PROJECT MANAGER

A project manager usually manages and/or coordinates the professional team, including the main contractor, subcontractors, engineers, quantity surveyor and even the architect. In my book, this is the most important role on your build after the architect: one individual who takes the responsibility for seeing the whole project through from start to finish. They can be a manager from your main contractor or an independent individual, but be warned: their powers can be limited if they are too independent because they may find themselves unable to direct the labour force on site unless it's via the bosses of the subcontracted trades. I personally recommend the management contracting arrangement, where you pay the project manager, you pay all the subcontractors direct at their trade price, but the project manager both hires and manages all the subbies. In other words, he works with people he knows really well and whom he can control on site, while you vest maximum responsibility in him.

BUILDER

You or your project manager will decide whether to appoint a main contractor (who can provide a 'turnkey' service, delivering a perfect finished and landscaped house if you want) or whether to package the work out to specialists. Typically, you might appoint a groundworks contractor for the foundations and drainage; a timber-frame company or block/brickwork contractor in conjunction with a carpentry firm for main construction; a roofer; a glazing firm; a landscaper; electrical and plumbing contractors; specialist engineers for reedbed systems or heat recovery; plasterers; decorators; joiners and fitters. Or you could just divide the construction between a groundworks contractor and a general builder. Or you could give the whole job to one firm while stipulating just one or two specialists that you want to employ. Your project manager may even be an employee of your main contractor. It's up to you. But I can give you three pieces of advice here:

Don't try to become a builder overnight and construct the whole darn thing yourself. I hope that around 60 *Grand Designs* programmes have proved that it is a miserable, slow and mire-strewn road to travel.

Remember that if your project manager works for your main contractor, you'll be hearing a selective version of events as they unfold. His loyalty will be first to his firm, second to you.

> I hope that around 60 *Grand Designs* programmes have proved that being your own builder is a miserable, slow and mire-strewn road to travel.

overleaf: **The finished result –
Sker House, in Wales, at the completion
of an epic conservation project.**

Don't be tempted to split the role of project manager between two site foremen or managers of two separate companies (such as the general builder and the timber-frame company). I have seen every arrangement of this type go wrong, with each party blaming the other for mistakes and a big gulf of responsibility between the two firms.

INTERIOR DESIGNER

If you're really posh you may feel that you need one of these. For that matter, you might think that interior design and layout is not your architect's strength, in which case a designer may help. They'll charge either on a time basis or a percentage of the cost of the work with which they are involved. An interior decorator will choose fabrics, curtains, wallpaper, carpet; they will often supply these products direct to you and take a cut from the supplier. Contact the British Interior Designers' Association (BIDA), via its website (www.bida.org), for a list of members and for the BIDA suggested scale of fees, which have been established in conjunction with the RIBA.

MECHANICAL AND ELECTRICAL CONSULTANT

Most people (and most project managers, for that matter) will appoint plumbers, electricians and specialist engineers as separate subcontractors, and then get them to work with each other to figure out how to put the engineering of the house together. The trouble is that homes are becoming more and more technically complex: we buy a German super-efficient boiler with European fittings and incomprehensible instructions, and then expect the local plumber to fit it. Or we ask the electrician to design our lighting system. Which is like asking a builder to design a house. Light and electricity are not the same things. Mechanical and electrical work is getting more and more complicated, with greater crossover between plumbing, electrics and specialist technologies. Groundsource heat pumps, biomass boilers, solar collectors, photovoltaic systems, sewage digesters, rainwater harvesting systems, air conditioning, underfloor

> Asking the electrician to design the lighting system is like asking a builder to design a house. Light and electricity are not the same things.

heating, heat exchangers, ventilation units, lighting systems and just good old-fashioned electrics and plumbing all need to be integrated in the modern house. More importantly, they need to be designed together to form one coherent environmental machine. If they're not, then the whole house could end up running grossly inefficiently. A mechanical and electrical engineer or contractor should be able to help you with a Schedule of Requirements, which for mechanical and electrical work is quite a specialized task, as well as a performance specification and a programme for installation. They'll also work closely with the architect and structural engineer to achieve the best performance out of the building and help you to specify products. That way, your home will be a beautiful, efficient and well-oiled machine, not some cranky Heath-Robinson crate. As with most professionals, their fees are normally a percentage of the construction cost of the mechanical and electrical services.

LIGHTING DESIGNER

As an ex-designer of lighting myself (I've designed schemes and fittings for nigh on 20 years), I couldn't miss out this category, even if the lighting designer of today is the equivalent of the feng shui consultant of 1995: a fad. However, some of my best friends are lighting designers and the brilliant ones like Bruce Munro, David Amos and Sally Storey of John Cullen are worth their weight in tungsten. As with all fashionable disciplines, the very best of the profession cannot be recommended highly enough and can literally transform your home, providing it with a chameleon-like ability to change character. The worst of them just bung in 50 low-voltage downlighters and then bung in a hefty bill.

USEFUL TIPS

· Good professional advice at an early stage can save money and time later. In particular, consult a quantity surveyor or structural engineer.
· Having employed an architect, take their advice on whether you need any other professional help.
· Carry out a proper assessment of your site at an early stage; this should include a soil test.

How to behave with your team

Some dos and don'ts: These are less advisory, more absolutely mandatory, if you don't want to end up as part of the foundations of your new home.

DON'T phone your architect over every little problem or concern. It's important to understand the roles of your professional advisors and the extent of their control and responsibilities. For example, an architect doesn't have direct control over who is on site when. It's generally up to the builder or project manager to control the site, the labour and the resources.

DO take advice from your professional advisors. If you've taken time and trouble to select them, trust them, listen to what they say and act on it.

DON'T give verbal instructions on site unless they've been agreed with your consultants and they're backed up in writing. If you just issue verbal instructions, you risk confusion and possible extra costs. If you're using an architect, the correct procedure is to give instructions through them and not direct to workmen on site.

DO take time to prepare detailed written briefs setting out your requirements for all your consultants. A good brief is an essential part of the design process.

Having employed an expensive team,
DON'T then try to run the project like some top-down Stalinist regime. Resist the temptation to seize control of the bunker.

DO allow your experts the freedom to do their jobs. Stand back from the project and enjoy your role as client: making decisions and hanging around, generally being a good egg and imbuing the world with optimism and excitement. These qualities are infectious and are also highly necessary for good morale on a building site. And do allow open, free communication between all members of the team. If you want to be kept abreast of developments, organize regular project meetings. During the build, these can occur once or even twice a week.

DON'T worry excessively, ask too many questions, look over everyone's shoulders, criticize or panic.

When on site,
DO say 'hello' and praise where praise is due. This helps to create the right productive atmosphere on the job. Do give positive feedback and even consider writing in a financial incentive for the builder in the contract.

DON'T listen to other people outside the professional team. You'll only end up getting very confused.

DO have faith not only in your professional advisors but also in the building team. This includes workmen from the most highly skilled craftsmen to a labourer. And do encourage everybody to work together. If there are tensions and rivalries between members of the building team, it's partly your job as client to help pour oil on troubled waters.

DO understand the roles and responsibilities of your professional team. This goes back to making sure that the appointments, responsibilities and lines of communication are all clear at the start of the job. I've seen many building projects where roles and responsibilities are unclear. This leads to confusion, mistakes and ultimately increased costs to you, the client.

When you're done,
DO ask everyone involved round for a party. Your home won't be just another day's work for them: the chances are that it'll have been an important part of their career. They'll be proud of it, want to see it finished and furnished, and want to take part in celebrating it. Moreover, given the team nature of any project, they'll have either formed or cemented working relationships while working on your build. If you ask any tradesman or professional advisor what they enjoy most about any project, their answer is nearly always the people involved – far more than the actual building.

Place
A plot on the landscape

When architects care passionately about landscape and context, they often end up producing a building that's an apology for itself: something 'invisible' or hidden, maybe with a grass roof or bark walls, so that it looks like a Teletubbies home or a giant molehill or a tree. The very best buildings that these architects produce often resemble large garden sheds. This can be something of a cop-out, given that great architecture should try to not only respond to, but also enhance, its setting.

At the other end of the scale marked modesty to arrogance are those buildings that shout 'look at me' and couldn't give a stuff about responding to where they are. These are often forbiddingly modernist glass boxes. I've noticed that architects may call this kind of building 'contextual' because it physically reflects its setting, like a giant 3-D mirror, which is also something of a cop-out when it comes to making a building look rooted in a place. I've also noticed that architects often refer to them not as houses but as 'pavilions', in the hope that we'll think it's some kind of temporary structure or garden folly. Nine times out of ten I think I'd prefer a shed, thank you.

Of course, there are always exceptions to prove the rule. There are some fabulous exemplary sheds, just as there are occasionally some beautiful glass pavilions, like Mies van der Rohe's Farnsworth House in Plano, Illinois (1945–51). There has to be room for every nuance and eccentricity on the scale of architectural propriety. But what really interests me is what lies between these dysfunctional extremes. I want less shyness and less arrogance and a lot more buildings that are socially adept and engaging and that charm me with a glow of radiant confidence. Above all, I want buildings that look as though they truly belong where they are. That means buildings that respond sensitively to their immediate environment and which look as though they couldn't fit anywhere else, at least not without a bit of a struggle. This is what separates proper architecture from sheds of all kinds, whether they're occupied by Grandpa or by Homebase.

Genius loci

Half of the design of your house will be about you; it'll be you-centric and you-stroking. The other half should be all about where it is: the landscape, the region, the immediate setting, the geology, topography and history of the place. I'm not suggesting for a moment that you have your house designed in any kind of traditional local vernacular, to look like an oast house or barn or wagon shed (though many local authority planning departments are keen on this reinventive, pastiche approach). I am suggesting that part of your brief to your architect is that the design should somehow embody the spirit of the place.

I want less shyness and less arrogance and a lot more buildings that are socially adept and engaging.

This sounds very wishy-washy, but it's not that difficult to achieve. The twenty-first-century architecture of Britain could be the most diverse in the world. That's because our landscape is so diverse. The fact that developers have been throwing up identical little Edgeorgian and Tudorbethan semis all over the country for the past 30 years is certainly a crime against humanity, but unforgivably so because Britain has a geology and topography that is immensely rich and inspirational and a history of localized building types that is varied and delightful.

There are two volumes that together illustrate this point magnificently. The first is the highly readable *Hidden Landscape* by Richard Fortey (Jonathan Cape, 1993), in which the geologist drives in his car from the Wash in East Anglia to the west coast of Wales, travelling, as he does so, through 1,000 million years of heavy rock evolution, from recent muds to ancient Devonian sandstones and granites.

On the way, his Morris Minor time machine gobbles its way through chalk cliffs, clays and flints, gravels, smooth sands, creamy oolitic limestones, hard crunchy carboniferous limestones, fullers' earth rock, iron stone, shale and, for afters, pudding stone. This gastronomy of rock represents a wider menu and a longer geological time span than you'll find in America, or almost anywhere on the planet. This myriad of textures and colours is our country's greatest hidden architectural secret and it is unique to us.

How this fabulous diversity of raw materials translates into our buildings is then explored in one of the classic reference works for the building enthusiast, Alec Clifton-Taylor's *The Pattern of English Buildings*, first published in 1962 (Batsford). In it he charts the way in which locally available stones, lime, clay and timber (which, of course, subject to geology, also varies in availability and species across the country) were historically translated into bricks, mortar, rubble and dressed building stone, roofing slate, beams and floorboards, providing each region and locality with its own sophisticated and often unique combination of styles and flavours. You can still see these differences as you drive through Britain, but you'll have to take A and B roads because our motorways are clogged with modern samey housing developments, like cholesterol sticking to arteries. Better still, take the train and look out of the window for half an hour and you'll notice how the pattern of traditional buildings varies as you travel. If you did that from a train window in America, you'd be forgiven for thinking that you were going round and round some infernal loop.

The reason that the style of our old buildings changes across the landscape is that building materials and their properties have always dictated how those materials should be used. Oolitic limestone, as found in Bath, is soft and could always be cheaply dressed to a smooth surface, whereas the carboniferous limestones of the Mendips or Peak District are much harder, leading to houses and farms being built of rubblestone in these areas. The length of hardwood beams and roofing rafters from oak and elm has always dictated the widths of rooms in simpler houses. And a building's roof pitch (one of its most characteristic features) was usually defined by the weight and precision of the materials used to clad it: lightweight, accurately cut slates meant tighter, more windproof joints, and so a shallower, flatter roofline was possible. Wobbly stone or clay tiles, on the other hand, required a steeper, and hence taller, pitch to prevent the wind from carrying rain underneath them, and to transfer their weight more directly onto the building's walls.

The local building dialects did get muddled, of course. With the spread of the canal system in the eighteenth century, manufactured materials such as clay tiles were transported far and wide. But generally, and especially in the case of lowlier buildings, the vocabulary stayed pretty localized because transport was such an expensive component of any heavy material. People shopped locally, but more often than not they also just chopped down a tree in their field, cleared some land of the stone, dug a lime pit or gathered some flints. A remarkable number of brick houses in Britain are made from bricks that were formed from clay dug on the site, out of the foundations, then fired in a makeshift kiln.

Reflecting a sense of place

This little historical/regional excursion serves one purpose. It reminds us that beneath the fatty layer of noddy houses and out-of-town retail sheds, it's still possible to identify the real grain of a place. This is useful when putting together ideas of what you want, and invaluably helpful when it comes to arguing the case for your design with your local planning authority. It is also never a bad thing to research a bit of local colour and flavour and incorporate some of these references into the brief you give your architect, especially if they're from outside your area.

facing page: **Hudson Featherstone's barn conversion near Norwich, Norfolk, won awards for the dialogue it struck between the new insertions and the vernacular character and materials of the old repaired building.**

He or she should, if they're any cop, design your house so that it fits its site like a glove. It should respond to the shifts in topography and vegetation and be orientated to maximize solar gain, views and privacy. And it should, if not copy, then certainly respond to the local building language. That response can, of course, be highly contemporary: a complementary response where the resulting house sits in direct contrast to the traditional language of the area, but a response where the argument between old and new ought almost to be explained in the new building as you look at it.

Alternatively, it might be a traditional response, where you copy or choose to incorporate some vernacular or local elements. Or, as is usually the case, it can be a moderated response, a design that might be contemporary in its massing, arrangement and answer to the geography of the place but which nods towards local building materials here and there. By way of example, compare the two contemporary Scottish houses, both *Grand Designs* projects and both from Glasgow practices. In Kilcreggan (see pages 155 and 168–9), Blast Architects chose zinc cladding as the roofing material, partly out of deference to the local tradition of using slate, but principally in order to mirror the colours of Loch Long over which the building looks. This is particularly important because the house – which juts out of the hillside – can only be seen from the road above, a view that sets its roof directly against the water. Near Stirling (see pages 73 and 158), KAP Architects chose to clad the house – itself a radical form in a traditional village – with split cedar shakes, in order that, from the road below, the house would appear to have been made from the trees in the woodland directly behind it.

So what happens if we don't design our houses to look as though they truly belong where they are?

They all, of course, end up looking like noddy houses: small semi-detached noddy houses or big executive noddy houses. Or the giant super-rich noddy mansions on gated private drives in Virginia Water or Knutsford or St George's Hill in Surrey. Very posh but still noddy. You can even order one out of a catalogue from one of the big national building firms as a 'bespoke' dwelling for your plot. But please don't do that.

Omi Architects' extension to a traditional building in Wetherby maximizes its glorious setting.

Place
A place on the planet

Time was, even five years ago, when eco-friendly living was a fringe activity for people with fringes who knitted their own sofas from organic porridge. 'Sustainability' was a nirvana where there were no cars and everyone grew their own furniture and heated their homes with shared bodily warmth. Now sustainability is a mainstream concept. It's no longer even necessarily dyed-in-the-organic-wool green, more a pleasing pistachio tint, which makes designing, constructing and living in an environmentally responsible house much, much easier than it used to be.

Admittedly, the word sustainable is now so overused that it no longer really means anything at all. It's been tried on by so many people, institutions and companies that it has stretched and gone all loose and floppy. Be warned that sustainability is now a big baggy sack into which people throw all kinds of old ideas and dodgy activities, just to get some eco cred, though you'll usually find the word just full of hot air.

Climatologists, however, and those close to government policy adopt 'carbon-neutral' as the Gold Standard of environmental performance, because it implies that whatever carbon dioxide is produced during any given human activity, somewhere along the line an equivalent amount has been sequestered away somewhere, probably in a tree. Some airlines already ask for a little 'carbon premium' on top of the ticket, so that they can plant trees to offset your flight to Barbados. Other organizations like Climate Care and The CarbonNeutral Company and co2balance will calculate your 'carbon footprint' on the planet and then plant trees to hide it, though this practice draws criticism from some quarters that carbon trading at this personal level does nothing to change our behaviour and make us less consuming of resources – the key problem.

As a result, most people are aware of the need to minimize our carbon footprint in order to minimize climate change. The WWF campaigns for one million truly sustainable homes in the UK because it sees how climate change is disastrously affecting the world's ecosystems and decimating the number of species on the planet. Its powerful model shows that at the present rate of consumption, we'll need three planets by 2050 to sustain our consumption of natural resources. Scary. Not surprisingly, its campaign is called One Planet Living and it calls for a more careful and responsible approach to construction.

Of all the terms available, from 'sustainable' to 'eco' to 'One Planet Living', I like 'environmentally responsible'. It's about the most honest and the most intelligent. I like it because it implies some degree of active engagement and preparedness to do something about the damage we're wreaking. Equally, the term 'environmentally aware' is about as weak and anodyne as you could possibly get, suggesting a state of sentience bordering on coma. While somewhere between the two lies the mediocre term 'environmentally sensitive', the sort of middle-of-the-road phrase that Marks & Spencer would think of to put on its plastic bags. This is the term that the Department for Communities and Local Government uses on its website to describe the environmental performance of its planned sustainable communities.

Historically, it was the DCLG's predecessor, the ODPM, that wrote the building regulations in this country, regulations that are becoming more and more stringent. Following the introduction of Part L in April 2002, governing conservation of fuel and power in a finished building, the government is now cranking up legislation. Part L will, each year or two, become more and more

> The WWF's powerful model shows that at the present rate of consumption, we'll need three planets by 2050 to sustain our consumption of natural resources. Scary.

demanding. At the time of writing, a planned Sustainable Building Code will require public sector buildings to comply with a new raft of standards, covering use of energy during construction, as well as during a building's day-to-day operations, and use of water. Within a few years this code will inevitably become standard in the private sector, and it'll carry on becoming more and more demanding. So why on earth, as a private owner, build your house to only just meet today's regulations? Why not plan ahead and make your home exemplary?

The UK's policy isn't an exercise in Labour government paternalism. It's being driven from Europe. And for that matter it's also being driven by consumer demand. There is a gentle culture change that is slowly happening in the UK, the slow adoption of the principles of popular sustainability. Most of us recycle; our local authorities have a mandate to recycle 25 per cent of our waste within a year or two; the combination of rises in fuel prices and the cheaper, more widespread availability of alternative heating technologies is driving more new house builders away from fossil fuels. And I can see among the projects I film an acceptance of alternative building technologies, such as high-performance 'engineered' wood products or sheep's wool insulation, not chosen on wholly 'eco' grounds but by normal people building normal houses with one eye on the building's performance and one eye on environmental issues they see covered in the news. In other words, green is no longer weird; it's now smart.

The impact of construction on the planet

Buildings on the planet are responsible for about 60 per cent of carbon dioxide produced. Housing alone produces more than a quarter of carbon dioxide in the UK and consumes more than half of our water. If we are to meet anything approaching the Kyoto targets, the way we construct and the way we heat our homes are going to have to change. According to the University of Picardie, in 2005 cement production alone was responsible for eight per cent of greenhouse gas emissions and was growing at a rate destined to outstrip aircraft emissions. For every tonne of cement made, a tonne of carbon dioxide is released into the atmosphere, half of it from the huge quantity of fuel required to heat the raw ingredients up to 1,500°C, and half of it from the ensuing chemical reaction.

In fact, in the UK we produce far less of the stuff than in other parts of the world. Spain is the largest consumer of cement, producing a tonne a year for every man, woman and child. Even Finland, Sweden and Norway produce more than we do – less than a fifth of a tonne per head. Moreover, Britain is a market leader in the use of alternative ingredients that pollute far less and use recycled materials: concretes that are made with large proportions of fly ash (from coal power stations and furnaces) or blast-furnace slag instead of cement. But these exciting PR claims are not reasons to be complacent. One study has shown that if every country in the world opted for the 'greenest' method of concrete manufacture, all we would do, given rising production, is arrest carbon dioxide outputs from cement manufacture at their current level: that's 1.8 billion tonnes of carbon dioxide a year.

Sustainable building in the twenty-first century

So how do we slow down all this environmental impact? How do we build more responsibly and move closer to One Planet Living? What is the ideal sustainable home to be building in the twenty-first century? Now that's a hard question to answer, because every material and every designed element in a building has an environmental cost and an environmental benefit. In assessing these, you have to look at a lot of variables and criteria. Some of which are overleaf:

above: **Studio KAP's house for the Leijser family near Stirling.**

Eco points to consider

1 Where materials have come from.

2 Whether they're recycled.

3 What pollution might have been produced in their manufacture.

4 What the embodied (sometimes called embedded) energy is in that material – something that's often down to shipping and delivery costs as much as the energy required in manufacturing.

5 Whether the materials are recyclable.

6 Whether further pollution will be produced in recycling (the release of chlorine, for example, from PVC).

7 Whether the product degrades or off-gases in use.

8 Whether the thermal or other performance of the product more than offsets any of the above values (a difficult one to assess).

And in terms of building design

1 Whether the design takes advantage of passive solar gain to heat the house.

2 Whether materials like concrete can be efficiently used as thermal stores (releasing stored solar heat slowly).

3 Whether elements of the design that have high environmental impact (such as a steel frame) are offset by performance advantages elsewhere (in terms of materials usage or thermal performance).

4 How efficiently the building is insulated.

5 Whether passive ventilation can be set up by using convection created by solar heat trapped in the building combined with venting systems.

6 The extent and efficiency to which carbon-neutral and energy-efficient technologies are incorporated (heat-recovery systems, groundsource heat pumps, wind turbines, photovoltaics and solar panels for hot water).

7 The efficiency of water management: waste water, greywater recycling, rainwater harvesting and economizing measures such as small-cistern WCs.

The point really is that the true carbon-neutral, environmentally invisible house doesn't exist, nor can it. Every human activity implies an impact on the environment and requires some energy – usually from an outside source like a power station or an internal combustion engine. Moreover, the chances of you or me converting to biodynamic eco-warriorhood are as likely as a 737 flying on biomethane from pig muck. I don't think I'm ever going to sell my house and give the proceeds to Greenpeace before building my own windmill-powered mud hut made from recycled compost and designing my own wooden car. Nor are you. And to be frank, the extreme examples of people who grow sustainable beards and eat their own cardboard sandals aren't going to persuade the rest of us to do the same, nor change the world. Composting toilets might well give you a feeling of closeness to nature: they're certainly very trendy among born-again greenies. But, as one water environmentalist told me recently, the methods that are used to process poo in big sewage treatment plants are identical to those used in domestic reedbed systems, and it's usually done on a far more effective scale.

A DREAM GREEN HOME?

I asked what the twenty-first-century sustainable home might be. The answer is, of course, lots of things. Elsewhere in this book I've argued for diversity, for a richness of building language that is so often missing in developer-built housing estates in Britain. So, I'm not going to prescribe what every house should look like, nor how it should be built. In the pantheon of eco houses, there is room for mud and adobe houses, timber frame, wattle and daub, as well as glass, brick, steel and concrete (especially when used in part earth-sheltered buildings for its thermal mass properties). A material's eco credentials depend entirely on use, context and clever design.

But if I were to put my money where my mouth is, lay the cards on the line and stick my neck out, what would I build? I've spent seven years following and studying housing projects of all kinds. I've visited hundreds and probably thousands of buildings. I've researched a whole host of building technologies and materials. I've even planned my own new build on the farm where I live, so I've given this question a lot of serious thought. So here goes:

- **With £250,000** and a plot ready to go, I'd start with the plinth – not beam and block but probably a thick solid one poured in limecrete (like concrete but made with lime, which is less polluting and more carbon-friendly) and its attendant clay-pellet insulating layer. Limecrete forms a super-insulating base for a house and has a high thermal mass, which means that it'll absorb heat slowly and release it slowly.
- **All the drainage** and pipework would be ceramic or metal. I'm not a fan of PVC because it's highly polluting to manufacture and leaches poisons as it degrades after use.
- **The structure would** be of timber panels – but super-engineered ones. I'm a fan of engineered timber, such as specialist insulating and construction boards, the best of which come from Germany and Scandinavia. Despite the presence of some glues and resins in these products, they have a pretty low embodied energy and they're very stable: they won't warp in Britain's moist atmosphere. And they're made from trees. Trees absorb vast quantities of carbon from the atmosphere, locking it into their timber. If we let trees die and decay, that carbon is eventually released, but if we chop them down and use the timber, the carbon stays in the wood. If we build houses from wood, we're locking carbon away, so we have a moral and ecological duty to chop trees down – provided, of course, that 1) we don't wreck habitats, 2) we plant and nurture more trees and 3) we use the timber to build with.
- **I'd probably want** to engineer in a bit of a steel frame as well, to provide some open-plan space and to minimize the need for too many internal rigidifying walls. But I might get away with using some engineered timber 'parallam' or 'gluelam' laminated beams here and there.
- **Insulation would be** sheep's wool, not because I get all excited about having skin-friendly home-grown fleecy wadding in my walls, but because wool is naturally hydrophilic, meaning that it absorbs moisture slowly and then releases it slowly, with a minimal loss of insulative value. This property is invaluable in conjunction with a timber frame. And there'd be more than double the quantity required by regulations to minimize heat loss.

- **I'd clad the house** in I don't know what: lime render, perhaps, or maybe a high-performing acrylic render, or timber or stone or … all depending where it was. Similarly, I have no view about the roof. Other than, were it to be flat, I'd use a PVC-free monoply membrane.
- **Glass might be** triple-glazed, but a similar kind of performance can be gained by using specialist double-glazing at much less cost. The frames, however, would have to be continental. We just can't make joinery like the Scandinavians can and I love those engineered softwood windows that often come with a super-performing powder-coated aluminium facing, made by Rationel and others. Top dollar.
- **I'd want to heat** my place in some kind of vaguely carbon-neutral way. As I write, we're installing a woodchip-burning boiler on our farm, which I'm intending to fuel entirely from coppice timber grown here. But I appreciate that that's not quite so practicable if you're building on an urban plot. Or, for that matter, if you want to heat less than, say, 1,400 square metres. The figures don't really add up. What are practicable, of course, are district heating systems using coppice fuel or woodchip or even grain as fuel, where several houses share one boiler between them. I can't for the life of me think why we haven't started to build housing developments in the UK heated in this way. They do it all over Europe.
- **Anyway, back to** my house. I'm hoping that with a south-facing glazing system and that limecrete plinth I'd get a fair bit of solar warmth trapped in the building, which I'd hope to augment with maybe a couple of solar panels for hot water and very possibly a ground-buried heat-source pump. These work like giant fridges in reverse, extracting solar heat from a kilometre of fluid-filled tubing buried in the soil over a huge area (up to quarter of a hectare – be warned!) and converting it through the miracle of latent heat (like a fridge) into warmth in the building, usually in the form of underfloor heating. Finally, one of the most persuasive technologies I've seen is a heat-recovery system. It's about the size and price of a boiler and works by extracting warm stale air from the rooms in your house while pumping in cool fresh air from outside, all with

an imperceptible quiet little fan. The really clever thing is that this machine reclaims up to 90 per cent of the heat from the stale air, warming the cool air as it comes in – so brilliant in concept, so simple in design and so effective at minimizing waste.

· **Now admittedly,** these alternative heating technologies might set me back some £25,000 if I adopted them all. I could pick and choose between them, perhaps, but ultimately they represent the biggest eco investment in my house. The jury's out on the payback costs for solar heating of water and photovoltaics (which still produce piddling amounts of electricity). But, as I write, oil prices are rising noticeably and there's talk of fuel shortages and problems with electricity generation over the next ten years. My instinct is to invest in being as self-sufficient and fuel-miserly as possible, by buying in the simplest technologies I can. I might even consider a domestic-size windmill at a couple of grand.

· **Finally, water would,** perhaps, be collected from the roof and stored underground for showers/garden irrigation/WCs. I just don't understand why, given that this country has less available piped water per head than Spain, we continue to clean, filter and chlorinate it just to flush gallons of it down the swannie every day. Greywater from baths would also go on the garden – perhaps, poo as well. I'm a convert to composting toilets, having used them more than once. They're easy to maintain, don't smell and produce beautiful powdery compost for your veg. Although I do completely admit that this last eco concession is born less out of my interest in sustainable construction and more out of my fascination with the veg garden (as a sometime grower) and the cyclical relationship it has historically had with the earth closet via the human digestive tract. Not a subject for polite conversation, but a personal interest of mine nevertheless. You could put in a nice white porcelain flushing loo with bleach in it and I wouldn't hold it against you. But me, I like the way in which the experience of using a compost dunny puts you back in touch with the earth. Every time you have a crap, you're giving something back to nature. Isn't that, um, beautiful?

facing page: **The Parkyns' Cornwall house is timber framed, insulated with reconstituted newspaper and fitted with argon-filled double glazing.**

A few practical eco pointers

First, here's a definition, commonly accepted, of what 'sustainable development' really means, viz: 'development that meets the needs of the present without compromising the ability of those in the future to meet their own needs'. Put simply, it's all about resources, how we manage them intelligently and how we need to adapt our lifestyles so that we consume less, consume it more efficiently and consume it with an awareness of its value. Sustainability is a key planning issue now, and any planning application should include a sustainability statement.

As to the design of your home (which may, of course, bear no relation to mine, see above), consider the following:

1 The building design should guard against heat losses and maximize solar gain. Any short-term environmental benefit of using 'eco materials' is a waste of time unless the building is highly efficient to heat and power: long-term is where the environmental impact of a building can really be measured. So incorporate ground-floor south-facing windows, the height of which should maximize the penetration of winter sun into your home; and put smaller windows to the north. Meanwhile, plan a thermal store in the building (a concrete plinth is particularly good at storing heat since it has a long 'thermal lapse': concrete and stone take a long time to heat up and cool down, unlike a timber frame, which has a disastrously short thermal lapse). The brilliant late architect Richard Paxton planned a long open-lap pool in his house as a thermal store, because water is particularly good at storing and regulating heat in a building

> Consider the polluting effects of materials, both in manufacture and disposal, not just the carbon-effect of a product.

2 Water consumption in the UK is rising but our supply is not, so we need to conserve water. Reduce your water use (and waste of hot and cold water) by installing aerating taps and showerheads, dual-flush toilets, self-closing taps and electronic sensor taps. You can also collect rainwater, then filter and store it for household use, and you can recycle greywater from baths, showers and basins, especially for your garden: it's much better for plants than tap water.

3 Materials that are used in construction should ideally be of low embodied energy, which is a term used to describe all the energy that's gone into their production and transportation. But before making a final decision, consider the polluting effects of materials, both in manufacture and disposal, not just the carbon-effect of a product. Where possible use local materials to reduce pollution from transport. Materials should also be non-toxic (which for me rules out PVC gutters and drains, most carpets and the use of tanalised treated wood anywhere it can get wet, allowing the arsenic treatment to leach out). Timber should be supplied by a Forest Stewardship Council (FSC) accredited source.

4 Use timber as your prime building material. Without doubt, timber is the greenest structural material (with the lowest embodied energy). From a design perspective it is a natural product that looks beautiful and can be easily worked. Added to which, using timber contributes towards carbon neutrality. This is because trees lock away carbon in their wood, which, if allowed to rot naturally over time on the forest floor, is released into the environment: construction locks away the carbon in the structure of a building. And thanks to modern breathable membranes and 'engineered' timber products like parallam beams, masonite and sophisticated hybrid panel products, timber construction today is a highly advanced technology. Wooden buildings are perfectly capable of outliving their masonry or concrete equivalents.

5 Be efficient with your design layout. A simple square box offers more internal space with less external surface than a long, thin design. A longer south face will, on the other hand, give much more scope for passive solar heating. Group your services together in the building for maximum efficiency. Put wall storage and bookshelves on the north, east and west walls to help reduce heat loss (and increase insulation). And think about a warm roof system, whereby the roof pitch is the insulated layer. This allows the roof void to be kept warm and used as habitable space. It also avoids the need to ventilate an attic space and so reduces draughts and therefore energy loss. Consider introducing buffer zones as well. A porch acts as an airlock, reducing ventilation heat loss and providing extra insulation. A single glazed conservatory on the west wall is helpful for clothes drying and can also harvest solar energy in the late afternoon, helping to delay the house's night-time cooling.

6 Minimize the potential for 'thermal bridges', in other words, weak spots in the external fabric where heat can track out, causing a thermal bridge. The most obvious points of heat loss are windows and doors, and single glazing is a fantastically efficient way of heating the planet, not your house. Even modern double-glazing presents a thermal weak spot in any building. And any metal structure, such as a window- or doorframe, must nowadays contain an insulating thermal break in its structure to prevent thermal bridging. Your architect will be aware of these problems and of the high-contrast issue of dew points and condensation, which can occur inside poorly designed walls and voids.

7 Reduce your consumption of electrical energy through design, but also through a careful selection of appliances. Buy A* or A-rated products like freezers and refrigerators and monitor how they're working. Thermostats have a habit of failing, often causing the unit to run day and night. Think about installing sun pipes, which transmit natural light into a house's interior and reduce the need for electric lights during the daytime. And naff as it may sound, use fluorescent lighting. I couldn't have recommended this technology even five years ago, but lamps have improved immeasurably in that time. At home I use temperature-balanced Dulux tubes (with a colour temperature of 2,900° Kelvin, to match the warm white of halogen lighting) and Philips Softone fluorescent bulbs, which seem to provide an ideal warm and friendly coloured light in table lamps. Don't buy just any old fluorescents because their lit colour will vary wildly, their light output will not be balanced (so coloured objects can appear odd under them) and they will have been designed for office work, where lighting is designed to be bright white or blue white (4,500–9,000° Kelvin). Always buy tubes that have the colour temperature marked, and generally go for anything between 2,500° and 3,500° Kelvin for use in the home.

facing page: **Alex Michaelis's sunken house in Notting Hill is environmentally highly performing.**

Place
A plan on the map

It is fair to say that although our planning system still functions, is still staffed by enthusiastic and visionary professionals and is being reviewed and overhauled almost continuously, we have, as a nation, lost faith in it. The Town and Country Planning Act 1947 (which still, incidentally, rather speciously carves major distinctions between those two places) is well past adolescence now. It hit a teenage wobble in the early 1960s and then a mid-life mini-crisis in the early 1990s recession. The real question is whether, after nearly 60 years, it's approaching retirement age. There's a lack of public confidence in the process itself, part of which is a mistrust of local government political intervention. There's also a frustration at the lack of cooperation from planning departments and a suspicion that legislation and centralized policy are too out of touch with local and modern needs. The whole structure seems to be getting slightly senile and incontinent. Maybe it's time to pension off the Act. If I could, I'd replace it with the Town and Country and Suburb and City and Village Design Act 2006.

facing page: **Pastiche doesn't have to be the default option – old and new in OMI Architects' extension to a house in Wetherby.**

The most common complaint I hear is, 'Why was my planning permission refused when my neighbour's wasn't and when, for that matter, Beano Homes got to put up 200 noddy houses down the road?' To which I have no answer. Save that 1) all planning is and should be contextual – to the extent that your neighbour might get the permission, you might not; 2) some planning officers are brave and mighty, some are cowardly and weedy; 3) there's no accounting for taste and much as you believe your scheme to be of the first architectural water, your local authority doesn't – and for that matter I agree with it.

As to Beano Homes' successes, get this. In a series of CABE (Commission for Architecture and the Built Environment) surveys, senior planning officers stated that the reason they let so many mediocre developments through was that they felt from experience that if they rejected them the decision would be reversed at appeal. Which sounds jaundiced and defeatist but reflects a lack of faith in the planning process from within.

How it works

The discipline we're talking about falls into three overlapping but distinct areas. First there's Planning, which involves the creation and implementation of policy. Then the ghastly monikered Development Control, which is all about doing just that (I'm on record as saying that if your council doesn't have a planning department but just Development Control, and you want to build something, go and live somewhere else). Then there's Urban Design, the new sexy kitten of the planning world that every young architect and landscape designer wants to get into, because it's all about remodelling townscapes with groovy modern architecture, water features and paving.

The truth is that of all three, the area to work in has to be Urban Design because it's creative, flexible and adventurous. And therein lies just one of the problems of traditional planning. Our system is weighed down by legislation and inflexible central government policy and a rigid appeals system. And it's a highly political one, reflecting, variously, the ambitions of national political parties and the often petty agendas of local government.

It's a contentious thing to suggest (and I've suggested it often), but the first thing we could do to improve the quality of our built environment is completely restructure the relationship between planning departments and local elected councillors – in effect doing away with planning committees, which can be highly political arenas, and introducing new advisory review panels made up of architects, local historians, building enthusiasts and so on. It's a brilliant idea.

The second thing we could do is put design (you know, that inventive, expressive, creative process that, apart from sex, men and women were put on the planet to do) right at the heart of urban, suburban and rural growth. Sounds odd that it isn't there already, doesn't it? But quality of design is not the average planning department's priority. The CABE surveys conducted between 2001 and 2003 showed a slight improvement, but in 2003 less than a third of departments that replied had a registered architect working in the planning department (32 per cent, down from 38 per cent in 2001); just a quarter use any kind of a design panel in their planning process; 80 per cent said that they could do with more design expertise and a witheringly tiny 15 per cent said that they used the magic combination of architect/landscape designer/urban designer in their assessments. When I tell you that a third of local authorities don't even have a conservation officer to look after their built heritage (listed buildings, conservation areas, high street frontages, old ladies) simply because it's not a statutory requirement, and so doing without just saves a bit more of the budget, it makes me want to cry.

Like architecture, in fact, planning needs to respond at a local, district and regional level.

Thirdly, we need to make planning local. Like architecture, in fact, planning needs to respond at a local, district and regional level. It isn't good enough that planners are just allowed to tinker with policy by writing their own development plans: policy itself ought to be generated out of the regions, not Whitehall. This change alone would itself inject enough design value into the process that the other changes might follow all by themselves.

How to work it

Of course, none of this is much use in helping you pick your way through the process of applying to build something. Or does it? Knowing your enemy allows you to exploit their weaknesses. Many departments are so understaffed and overburdened by their obligations to meet government targets (for processing a certain number of applications within the eight-week statutory period) that junior officers are often led to reject applications on technicalities. It might be a dangerous move, but you could consider the tactic of whipping up a little friendly local opposition to your proposal (as well as maybe some letters of support). This, at least, has the potential effect of kicking the application up the administrative ladder by making a song and dance about it. I even know of one local authority where builders wait for one officer in particular to go on holiday before putting in their applications.

Just as there are technical grounds on which an application can be rejected (not showing your workings-out, not putting your name at the top of the paper), so there are techy boxes to tick as well. Providing a proper design statement with your application, for example, or an effective sustainability statement. Or, for as long as planning committees are composed of lay councillors, providing drawings that aren't just technical but which are easily readable, nay, beautiful.

> One of the most tactical devices I've come across is to incorporate sacrificial elements in your design: components that you are prepared to negotiate away.

One of the most tactical devices I've come across is to incorporate sacrificial elements in your design: components that you are prepared to negotiate away but which you don't really want anyway, leaving you, hopefully, with exactly what you wanted in the beginning. It's the art of negotiation, isn't it, when both parties come out of the talks thinking they've got precisely what they wanted? One couple of architects, whom I'm not prepared to name, did include a clutch of sacrificial components (windy staircase, floating walls, experimental lookout tower) in their planning application, but did such a good PR job on their neighbours, councillors and planning officer that, in the end, their entire proposal was passed without condition. So oh dear! They had to build the house as drawn: floating walls, lookout tower and all. Be warned.

But you may be lucky enough to have an inspired and engaging department where you live – planning officers who'll come out and look at your site and make a positive contribution to your ideas; a committee of passionate politicians who champion good design. Don't laugh; it does happen, and surprisingly often. Despite the system, many local authorities are remarkably enlightened and seem to be taking the responsibility for policy into their own hands: Woking has one of the most advanced sustainability agendas in Britain; Bexley claims to put design right at the heart of its process, and there are many others across the country. One planner wrote to me recently saying that he felt that the best buildings resulted from the collaboration of four people: a gifted architect, a visionary client, a talented builder and … a brilliant planner, of course. He didn't tell me which authority he works for, but if I can find out I might have to move there.

2 Dreaming

Dreaming
Daring to dream

The biggest risk for any architect is when he or she comes to actually build something. Entire architectural movements have risen on, and been sustained by, an oeuvre consisting of completely theoretical buildings, some of which would have been highly undesirable had they been built, and most of which were technically unfeasible anyway. The safest kind of architect is an academic one, a theoretical one, because their mettle can never be truly tested. All praise, then, to those practitioners who not only design houses for real clients living in the real world, but who also produce, against all the odds, really fantastic buildings.

The design process follows pretty well the same route, whether for a spoon or a concert hall. It starts with a brief and a period of enquiry. What might follow are bursts of vivid imagination in the designer's mind, some maturing, some reflection and general picking about of ideas and then, in summary, the tentative/assertive statement of those ideas. And that just gets you to the proposal stage, which is easier than any other because it usually involves only a few individuals. When design starts to get really messy is when loads of human beings get involved.

So the design for your home is probably at its most perfect when it's still on the drawing board. On presenting his ideas for a house to his client, an architect was flattered to be told how beautiful the house was. His reply was, 'This isn't a house, it's a piece of paper with some marks on it.' Indeed, many architects are wistful about drawings and see the whole building process as a series of inevitable compromises that have to be struck just to get a building up: compromises with planners, owners, builders, accountants. This is sometimes referred to as 'the continuing design process' and the secret, one architect has told me, is to fight your corner till the last and compromise on the things that matter least to the integrity of the design but most to everybody else.

Think on this: at least the spoon designer gets to make a prototype and try it out. He can change and refine it, whereas the architect has to live with his prototypes. There's very rarely a chance to go back and remodel something; every project (worth its salt, frankly) is a first, and so a source of potential error or even a series of errors. As Frank Lloyd Wright said, 'A doctor can bury his mistakes, an architect can only advise his clients to plant vines.' It all makes you want to go and buy an executive home from a catalogue, doesn't it?

But a house is not a spoon. It's a much more sophisticated piece of engineering made from thousands of pieces that nowadays have to be highly engineered to meet stringent regulations and tolerances – or at least it should be. Most houses are still constructed by nailing sticks of wood to other sticks of wood, gluing hundreds of bits of cooked earth together, wading around in wet grey goo and drinking cups of tea. A house should instead be manufactured off site, in a warm factory by skilled enthusiasts – and one day they all will be. Putting up a house on a building plot is like trying to build a prototype car from scratch in the middle of a field. It's a miracle that anything decent ever gets constructed. But it does.

> Putting a house up on a building plot is like trying to build a prototype car from scratch in the middle of a field.

What follows in this part of the book is a highly personal choice – my choice – of miracle buildings that I think are seminal, exciting and exemplary, that celebrate the great triumphs of good design over adversity. They're divided up into typology: houses and apartments in towns and cities are usually designed and built with a whole range of pressures and exigencies that are different from those that apply in open countryside or in villages, so each type is treated separately.

previous page: **Noel Wright's house for the Parkyn family in Cornwall.**
overleaf: **Ken Shuttleworth, founder of MAKE Architects, designed his own uncompromising and beautiful house in Wiltshire.**

There is no accounting for taste, of course. That's why car manufacturers offer you just a certain choice of car colour and upholstery trim, but not in any combination you like. They don't want you driving around in something they'd be embarrassed by. It's also why we have Planning and Development Control, local government departments that are there to regulate and improve our planning applications and instead assert their taste. For once, though, I'm able to get my

own back. You may not agree with me. None of these projects may correspond to your dream or be appropriate solutions for your site (after all, the influences of people and place form the overriding distinction between just any old building and proper architecture). You might even find the range of my choice rather limited. But if you don't like any of them, you probably don't like what I have to say, either, and so shouldn't have bought this book in the first place.

New Urban

It is now very fashionable to live in cities in Britain.
That's not something you could have said in 1995, or even
in 2000. Now, architects like Ken Shuttleworth and Ian
Simpson are designing the new generation of glass-clad
super-towers, for entire communities of young urban
singles and childless couples who can enjoy the glamour
of urban living, urban shopping and urban drinking in
places like Manchester, Leeds and Birmingham (yes,
Birmingham, which has undergone an extraordinary
transformation). But fashion aside, city living has a lot of
appeal. For a start, we are social creatures. On the whole,
we like other human company and tend to thrive in tribes.

And although you might argue that the maximum efficient size for a human tribe is about the size of a village, I'd argue that two of Britain's most vibrant and successful cities (London and Sheffield) are actually a series of interconnected villages, each with an independent character and culture.

Part of the renaissance of central city living is to do with the way we are trying to subjugate the car to the pedestrian and give our cities back to people. Partly it's because of the inevitable property cycle, as young impecunious professionals look for cheap accommodation in areas (Shoreditch in the 1990s, for example) that then grow economically and become fashionable. I think it may also be to do with the way that Country Living, the great get-away-from-it-all philosophy of the 1970s and 1980s, isn't all it's cracked up to be. Not in February, anyway. Try getting a frappuccino and a conversation about something other than schools or field sports in deepest Mudshire. OK, you can get the coffee nowadays, but the social life is never going to be as vibrant.

There is so much going on in our cities. So much to absorb, adapt to, connect to. When it comes to how our cities are designed, the same is true. So much urban

above, previous and facing page:
**Monty Ravenscroft and Claire Loewe squeezed
their top-lit skinny house onto an unpromising
site between two nineteenth-century houses
in Peckham, London.**

design is about fitting in to what already exists. Lesser people might consider this a disadvantage. For those steeped in the spirit of urban life, however, this need to be responsive can draw its energy from the very exigencies that limit a project.

I filmed Monty Ravenscroft and his partner Claire Loewe build an extraordinary, experimental house in 2004, on an unpromising site in south London (see pages 97–9 and 110–13). Designed by the late, brilliant and sadly missed architect Richard Paxton, but engineered by the polymath Monty, the house effectively had to be hidden from the street and not impede light from reaching the adjoining houses. So it was low, flat-roofed, cranked to follow the asymmetrical site and lit by skylights. In deference to its urban setting, it also employed alternative high-tech construction techniques borrowed from commercial construction. A super-lightweight steel frame and stressed plywood skin made the building a super-insulated mobile home without the wheels, one that was also minutely tailored to the site.

> In a sense, the Anderson House came about from the most extreme tactic available in a high-pressure urban environment: if you can't build up, then build down.

Sometimes, no matter how close you build to the boundary, a site is just not big enough. The Anderson House in London (see opposite and pages 116–19), designed by Jamie Fobert for David Anderson, would just not have been a viable proposition had the architect not magicked some extra space from somewhere. The brilliant solution won the architect the 2003 Manser Medal. The site had been a land-locked bakery and even now has no façade: it's not a building you can photograph from the outside. In a sense, this house came about from the most extreme tactic available in a high-pressure urban environment: if you can't build up, then build down. This house needed 18 concrete piles, a steel frame, 400 drawings and 26 party wall agreements for it to exist. The clever thing is that for the most part it doesn't seem to exist. Like Monty's house, it capitalizes on modern commercial construction techniques to create new and stimulating environments. The construction is in concrete, which can be very polluting to manufacture, but the structure is robust and proof to the vagaries of taste because it's invisible. The place works like a cave, with a regulated temperature all year round and minimum heat loss. I don't think we build underground enough in this country.

facing page: **A skylight in Jamie Fobert's hidden Anderson House.**

Nor do we, really, take advantage creatively of the haphazard nature of our cities. Monty has built his house on a tiny strip of land that was, apparently, useless for anything but garages. The Anderson House reuses, or started with, a small commercial site. The Voss Street House, by the architect Sarah Featherstone, brilliantly does both. The house (see opposite and pages 120–1) sits on what was a typical traditional urban site: among a row of shops, three storeys high, with grotty sheds, workshops and a mews behind. Featherstone's house doesn't really replace these so much as knit itself into them. You enter the building from the mews, as if into a garage or shed. You circulate round a tiny courtyard garden and the rooms spill off a winding stair-tower, forwards and backwards, tucked above the garage, behind the shop and, for that matter, over it. The house occupies all the space that a shop owner wouldn't know what to do with: that inaccessible third floor, the useless yard and leaking outbuildings. These are exploited and their inadequacies ingeniously turned to exciting advantage. That's the kind of thing you pay an architect for. That's why they train for seven years.

Of course, occasionally they get to design something new, which doesn't involve inserting an invisible building under, over or through something that already exists. The house that Jeremy Till and Sarah Wigglesworth built in Camden, by the Great North rail route into King's Cross, was constructed on the site of an old forge but is brand new (see overleaf and pages 122–5). This would pose any architect the conundrum of how to fit the style and language of the building into what is always going to be a highly textured context, in this case a mixture of old and new housing plus the infrastructure-scape of railways and overhead power cables. Being both very sophisticated architects (Jeremy is a professor at Sheffield University while Sarah's practice is considered to be among the most progressive and enquiring in Britain), their response was also sophisticated, what Jeremy calls a 'hairy' building, bursting with detail, symphonically articulated and full of prodding, investigative ideas. It's a brilliant house, but clearly not one that could be or should be just reproduced, because it's almost impossible to imagine it anywhere else. It's a house that attempts to reproduce the diversity and haphazardness of everything around it, that isn't so much a building as a piece of urban landscape in its own right.

facing page: **The newly modelled Voss Street House by Sarah Featherstone overlooks its own enclosed private mini-yard.**

The Stealth House (on pages 108 and 126–9) is, however, something altogether more obvious and coherent, 'something uncompromisingly modern but with urban manners', as its architect Robert Dye explains. He designed it for the comedienne and author Jenny Eclair and her partner Geof Powell and it won the 2005 Manser Medal. Geof wanted to self-build and manage the project, so it was essential that the design was easily constructible. In fact, what looks like a complex structure with openings that move in and out of the surface is actually a straightforward timber-frame box. That and the high levels of insulation in the building ensure its environmental credentials, as well. Geof wanted a beach house but was prevented from living next to the sea by Jenny and their daughter, who couldn't leave London. So Geof built a beach house anyway, with a one-metre

shingle beach next to it, just for good measure, and although the timber cladding might be redolent of tar-painted sail lofts, in fact, this building borrows all its language from its immediate surroundings: from a pre-war moderne house three doors down, from black-painted cladding on the 1960s block opposite, and from the pitched roofs of the neighbouring Edwardian terraces. But, in fact, it isn't borrowing. The design selectively absorbing and restating – with 'urban manners' – its context.

There are a million ways in which a building can reflect its environment and its immediate setting, using a million stylistic devices and a million materials. You could argue that given the denser, richer textures of the cityscape, the choice there is greater than in the countryside where the conventional palette of grass, bark and mud seems rather limiting. The mix of historical styles also tends to be denser in the city, and often jumbled, too, camouflaged by the way buildings

above: **Jeremy Till and Sarah Wigglesworth's home and studio in north London.**

get adapted, extended and built over. Look at Henning Stummel's eye-popping addition overleaf for a truly twenty-first century example of the 'extension'. This is a vertical, multistorey bathroom and loo pod; an honest and highly functional addition to a traditional high-rise town house that doesn't waste an inch of space and yet brings the full sybaritic experience of modern ablutions to a period home. For that reason alone it's a clever piece of design, but I particularly like the outside treatment of the building, which is coursed, like brickwork, but worked in the complementary materials of wood and frosted glass. It's a half-timbered spaceship – literally.

It's absolutely appropriate that when we add our modern topcoat to the rich layer cake that is our built environment, it should be in a modern style. Cities get built like super-fast geological strata, each seam representing maybe just 20 years of building. And it isn't just buildings and infrastructure that contribute. The commercial pressure that gets placed on the high street adds an extra layer of design, much of it at odds with the buildings it cloaks. But this chaotic mix of hoardings, street signs, historical building styles, neon and tarmac is but a feast for the eyes of the urban designer. Many architects revel in the complexity of the mix and the way that, in the overlayering of ideas, some resonate with each other and survive the passage of time. Sean Griffiths is one of those architects. His partner Lynn said of their own home (on pages 109 and 114–15): 'We saw an eighteenth-century building further up the road that acted as a billboard. It advertised plywood with large text made from plywood. It announced what the business did on its façade. We thought, "Why can't we make a building like this?" We wanted to make a delightful surprise, not something bland.'

Griffiths's work certainly isn't that. I remember a design that he once pitched at a meeting of borough councillors who had the thorny and expensive problem of reuniting two halves of a town split by a very busy main road. His solution wasn't a bypass or an underpass or a tunnel or a flyover. He just came up with the idea of a zebra crossing two hundred yards long. Shame they didn't get it.

It's absolutely appropriate that when we add our modern topcoat to the rich layer cake that is our built environment, it should be in a modern style.

overleaf: **Henning Stummel's extraordinary spaceship extension to a London town house accommodates functional ablution rooms.**

Griffiths's practice, FAT (Fashion Architecture Taste) specializes in what some people think of as pastiche, others as postmodern wryness. I think of it as resonant architecture, because FAT's work has this ability to communicate all kinds of ideas about place, history and use to all kinds of people in very easy, enjoyable ways. Griffiths's own house is like this: fun, delightful, alternative. It's not the kind of architecture you want to see plastered all over the country but, in the name of diversity, every community in Britain needs, somewhere, something along the same lines, because this kind of design is so accessible and human.

One thing is absolutely clear about all of the projects featured in this urban chapter: the fact that in the city or town, architecture and design can never work in isolation. The environment is too pressurized for it to be otherwise. The countryside offers the luxury of space and even a sense of isolation. The urbanscape, however, is all cheek by jowl, built up, higgledy-piggledy and full of what urban designers like to call 'grain' – grain that requires urban buildings to respond and flex in their appearance. That's not to say that you should design your house to be a jumble. The magic of architecture lies in creating order out of and within chaos, and in making space from nowhere.

above: **Jenny Eclair and Geof Powell's Stealth House, designed by Robert Dye, was the 2005 Manser Medal winner.**
facing page: **Sean Griffiths of FAT designed his own elaborately faced house in London.**

Monty Ravenscroft never relaxes, except to look through his open skylight – a giant connection to infinity. The home he built is an exercise in squeezing – squeezing space out of nowhere; 'squeezing' the building in the middle of the site to give it a 'waist', so that it fits its location; and squeezing perfection out of dogged dullness. Not surprisingly, much of the wall space is given over to storage, particularly since there are no windows in most of the building. All of the cupboards lean, float or perform geometrical tricks to conjure the illusion of movement and space.

FAT House
London

Sean Griffiths is one of architecture's idiosyncrasies, a man whose practice (Fashion, Architecture, Taste) is always looking for extra resonance in places and in popular culture. His own house, though fairly conventional and modest inside, takes on its surroundings by the short and curlies, mixing local industrial vernacular with cartoon language (the 'little house' cut-out) and serious stylistic language from the architectural histories of Dutch and American Modernism. You can't say it's not eclectic.

Anderson House
London

The Anderson House, like Monty Ravenscroft's (pages 110–13), cheats space, light and view out of the most unpromising site. The house is squeezed between and under surrounding buildings so that it is utterly camouflaged. All you see is one window and a front door. Windows aren't windows, but skylights, lightwells and glazed openings punched through unexpected bits of building. The resulting shapes are exciting and bring light deep underground.

Voss Street House
London

Sarah Featherstone's house in
east London is a glamorous new
build 'stepped' through a series of
older buildings, tailored into its
site seamlessly. Rooms are well
lit and private, facing their own
internal courtyard, a reinvention
of the backyard that had been
there. A spiral staircase tower is
the spine of the building. As
rooms connect and rise through
the house, some of those
connections are open-plan.

Sarah Wigglesworth and Jeremy Till's home is unusually expansive but explorative. It uses all kinds of devices and quirks to 'knit' itself into its setting. The building employs several separate building techniques – steel framework, timber panelling and a bedroom wing constructed with straw bales over two storeys: acoustically ideal next to a mainline railway. The office wing (opposite and overleaf) has stone gabion supports. The interior (right) is as dynamic and experimental as the exterior.

overleaf: **Jeremy and Sarah by their stone gabion supports for the office wing, and their 'honesty panel' in the straw bale bedroom wing.**

Stealth House
London

Jenny Eclair and Geof Powell's
house, designed by Robert Dye,
is a beach house in south London
– without the beach. Inside and
outside there are references to
the building's context and the
surrounding houses: black-clad
1960s housing blocks and
Edwardian brick terraces.
The house isn't large but Dye
has performed the expected
architectural trick of magicking
space from nowhere.

This place is a lesson in how to use materials. In other words, expediently, frugally and sensitively. The palette and the budget were modest, but the resulting sensation is of luxury. It's all down to how you exercise your eye, not your wallet.

New Suburban

Most people, myself included, who got a whiff of suburban living in their youth will have since spent a good portion of their daily energy trying to escape it. For us, suburbia is synonymous with the dead end of the road; it's a claustrophobic, stagnant backwater of polite middle-class living: manicured, smug and comfy. But not everybody feels this way. My wife, who unlike me has never been encumbered with having to rise through the mid-twentieth-century class system, has a fascination for suburbia, harbouring deep-seated yearnings to live there. Mingled with the smell of lawnmower petrol and Glade air freshener, she can detect more exotic scents: those of containment and contentedness.

Compared with where we live now, in the middle of deepest Mudshire, there are also lots of practical advantages. Suburban children can walk to school or get the bus, and can visit friends without their parents having to spend their weekends minicabbing them around. Suburban houses are generally easier to maintain, weed, mow and keep tidy than some overgrown, sprawling, rural mess in the countryside. And they can be contained, tidy spaces: somewhere to marshal and order your lives. This holds a lot of appeal for my wife, the busy adjutant to our sprawling, rural and messy household. Even I can see the attraction.

Architecturally, the suburban glades of Arcadia have much to offer. Cityscapes may be high density, high rise and high grained. Rural design tends to be open, adventurous and often egotistic. Suburbia comes halfway between: the best suburban architecture is medium density, textured to avoid bland repetition, and polite. It's polite to its neighbours, polite to nature, with which it dances a very formal and manicured pavane, and polite to the rest of the world. You might find politeness overrated and want your design to be in-yer-face subversive. In which case I'll be just as rude back and tell you to bugger off back to the town, or build in somewhere completely isolated and remote. Politeness is a code that allows group behaviour. It's the lubrication of human group life and much underrated as a social tool. And when human beings are corralled into housing developments, politeness becomes just as necessary in architecture as in everyday social engagement.

There are plenty of polite – and interesting – housing developments from the early years of suburbia: Welwyn Garden City, for example, or Surrey, which isn't a county, more of a series of parks interspersed with posh, polite houses – so much so that Noël Coward referred to its countryside as the 'Tames of Surrey'. Every regional town has a well-groomed 1950s or 1960s housing estate of pitched-roof modernist semis. But it's hard to find the twenty-first century equivalent. So I've included only two here: one is the Bloembollenhof housing estate in Vijfhuizen, Holland, by S333 Architecture + Urbanism, exemplary for both its architecture and its social planning (see opposite, and pages 131 and 142–3). The buildings are obviously house-shaped, but their detailing and simplicity is just as redolent of the other kind of suburban buildings: the sheds and barns of light

previous and facing page: **Bloembollenhof housing scheme in Holland.**

industry or market gardening. And although there are only four basic forms used here, there are 52 different shapes of home, which allows residents to customize and adapt their house. More importantly, this flexibility means that some can be large single dwellings with double-height spaces while others are more modest social-housing homes. Yet externally they appear similar: modestly architecturally interesting and, of course, polite. People of all means and backgrounds are encouraged to rub along here, to share their open spaces and celebrate what

they collectively own, which is a strongly branded, exciting community.

The other scheme is a little less restrained. It's Abode, designed by Proctor & Matthews for Copthorn Homes and New Hall Projects, in Harlow, Essex (see left and pages 144–5). This is part of a much larger development, which in turn is part of the current massive expansion of Harlow, but it retains all the diversity and architectural interest of the small-scale or one-off project. The Commission for Architecture and the Built Environment has described it as 'a vibrant mixture of bright colours, crisp contemporary detailing and traditional materials such as shiplap boarding, leadwork and thatch'. All without the slightest hint of twee. It remains to be seen whether the current fashion for this hybrid mix of contemporary and vernacular design lasts beyond the flash in the pan. I hope so. These buildings manage to look both utterly of their time and, importantly, utterly of their place. That's a hard trick to pull off but such an important one: as I say in 'A Place on the Planet' (page 70), if any of the hundreds of new communities and millions of homes that we need to build are going to be

above: **Abode, in Harlow, Essex.**
facing page: **A remodelled 1960s detached house by Jeremy Spratley.**

at all sustainable, they first need to appear as though they belong where they are.

Both of these developments rewrite the old language of suburban but in a modern idiom. The same can be said of Jeremy Spratley's one-off remodelling of the Evans family's house in Marlow, Buckinghamshire (see pages 135 and 146–7). The interior of this 1960s home was gutted and opened up to form larger and open-plan spaces, then refloored, reclad and rerendered. Think of it as a chassis-up rebuild of a 1964 Mini Traveller, but now with an E-Type engine and sat-nav for good measure. With all the fun and good points of 1960s suburban living and none of the inconveniences, it's a favourite project of mine – not least because it reminds me of the house I grew up in.

The pair of Nail Houses by Glas Architects (see opposite and pages 148–9) is another rewriting of the same language, a reworking rather than a remodelling, because these houses are brand new, in fresh brick and glass, but they couldn't exist but for their mid-twentieth century antecedents. Pure evolution.

But there's room for some new revolutionary voices among the leafy groves of suburbia. Not content with reworking pointy roofs and casement windows, some adventurous architects have broken the traditional suburban mould and forged a new language that, at its best, is highly contextual and site-specific. Their houses, as a result, are one-off projects that respond to their immediate location – to the semi-tame natural environment around them and to the other buildings in the neighbourhood. This is very flexible architecture, sexy suburban, sinuous and accommodating, and we need more of it.

> This is very flexible architecture, sexy suburban, sinuous and accommodating and we need more of it.

There are two examples here. One is Anjana and David Devoy's curved glass house in Clapham (filmed for *Grand Designs*), designed by Peter Romaniuk and Jack Hosea (see overleaf and pages 150–1). This house responded to its built and natural environments very literally, by filling its rectangular semi-urban plot on three sides while curving around a huge chestnut tree on the fourth. This shape (in plan a rectangle with a scoop carved out of it) may be simple and the treatment of the exterior (in wood and render) may be straightforward, but together they evince the delicate dialogue that is played out in suburbia between

facing page: **The semi-detached Nail Houses by Glas Architects.**

the man-made and the natural. A dialogue that, in terms of this building, is resolved in the beautiful craftsmanship of the woodwork. Tree makes wood, wood makes house, house thanks tree.

This kind of poetic relationship is just the kind of thing that is diffused through the work of John Tuomey and Sheila O'Donnell. Their architecture is so lyrical it can almost be described as built poetry. And nowhere is it more elegantly expressed than in Howth House, on the shores of Dublin Bay (see opposite and pages 152–3). As in Clapham, the overall form is a response to the house's immediate semi-urban setting: the building's plan relates it to the surrounding buildings, the nearby street and the vista, which means that it takes on a kinked shape as it reorientates the visitor's view out towards the sea. And, as in Clapham, this house uses a soft, friendly material, a grey lime render, to assuage the brutality of its form and relate the surface of the building to the one great natural presence that it points to: the sea.

A good architectural experience, a sense of place and poetry are all things that have been missing from mass housing for decades, which is a crying shame. Come to that, they are rare even among one-off commissions, which is verging on the criminal. According to think-tanks and trend gurus, suburbia is currently the place to move to; it's become cool. If we're going to build there, we've got to build more than just cool architecture. It's also got to be contextual, lyrical and, of course, polite.

previous page: **The Devoys' house in Clapham.**
facing page: **Howth House on Dublin Bay,**
designed by Tuomey and O'Donnell.

Bloembollenhof Estate
Vijfhuizen, Holland

Bloembollenhof is a progressive community architecture project in Holland. Many are surprised that such a socially successful scheme should look so severe, but these democratic façades hide a mix of social housing and privately owned dwellings in a boggling array of sizes, from one-bed apartments to large family houses. The architects S333 have deliberately cloaked all these social differences in a unity of style and a rigid design code that I find curiously pleasing. A breath of fresh Dutch Calvinist air.

Abode
Harlow, Essex

This housing scheme by Proctor & Matthews Architects for Copthorn Homes is a really worthy stab at creating a place from nowhere. Traditional and local materials, like brick boarding and thatch, are absorbed into a busy and varied palette of materials, shapes and textures that attempts to brand itself as Harlow's 'new vernacular'. Here is an exciting alternative to the sea of anonymous noddy homes that have sprung up in Britain since the 1970s – we need more projects with this level of interest.

Rather than move, the Evans
family chose to stay in their 1960s
house in Marlow and employ
Jeremy Spratley to remodel their
house inside and out. I've put
this project here to prove that
the 1960s detached and semi-
detached houses are not dead –
but perfectly capable of being
upgraded and extended. Now
the Evans have the three
requirements of the modern
mantra: space, light and a
connection with the outdoors.

Nail Houses
London

If 1960s housing can be revitalized (see pages 146–7), then the semi-detached house, a model that's been popular since the turn of the twentieth century, perhaps has an unlimited lifespan. Glas Architects' Nail Houses are the pared-back super-modern version. It's as though they designed a cool brick box, punctured with openings that slide in and out of the building, and then realized that if they stuck a mirror-image of the house right next door, people would find it less forbidding and much more familiar. They were right. This is uncompromising modernism – tough and powerful on the outside, wonderful on the inside, and delivered in a friendly package.

Curved House
London

Anjana and David Devoy's unusual home in Clapham revolves around a chestnut tree, and so derives its meaning from the way it turns its back on the surrounding estates and streets with a wall of concrete, while embracing a pocket of suburban nature with a timber façade. It makes use of the no-man's land space between streets at the bottom of the suburban garden and consequently captures views, privacy and light by looking right along the line of neighbours' garden fences, as well as drilling down to exploit the space a basement can provide.

Howth House
Dublin Bay

Suburbia sometimes offers extraordinary delights and possibilities. Howth House in Dublin Bay by John Tuomey and Sheila O'Donnell cranks itself across a dog-leg site and orientates its occupants to drink in the views, like an enormous viewfinder. All eyes are on the sea, whereas surrounding buildings are cleverly framed out. It may be a ruthless, even antisocial tactic, but it delivers a slice of perfection: succour to detox us from the day-to-day ravages of the world.

New Rural

I can't exactly figure out why so many people want to
live in the country. True that, unlike any other country
in Europe, we have a history of the aristocracy spurning
the smelly city at the earliest opportunity and heading
for the hills. So, given that Britain has also historically
been class obsessed, my conclusion is that we're all trying
to be nobs. Which would neatly tie in with our obsession
for swanky 4x4s. Or perhaps we are, underneath it all,
a nation of pantheists, finding our Protestant God in the
woods and the groves. Despite our famous shopkeeper
pragmatism, the Romantic Movement took firm hold
in Britain in the nineteenth century: Wordsworth and
Coleridge believed that their new-found faith in nature
could change the world.

The reason I live in the middle of Nowhere-on-the-Hill is so that I don't have to spend 45 minutes driving three miles through solid traffic to my children's school in the morning (though I do have to drive them everywhere at weekends). And I can laze around in my garden without some busybody neighbour next door telling me to cut my hedge. And I can see the stars at night. As a fan of Wordsworth, I thank my lucky cluster that I can see a sunset and sunrise, or catch sight of a buzzard or barn owl in the lane, and get my hands dirty in the soil without worrying that I'll accidentally spike myself on a discarded syringe. Like so many post-industrial refugees, I have spurned the city to find a place where I can feel a sense of place, belong and feel safe. It's that simple.

Of course, this means living in an old house because there's precious little good new architecture in the countryside. To find great rural buildings, you need to go to Arizona or Switzerland or Japan or New Zealand. Some countries specialize in exciting new rural houses. You might think, from watching *Grand Designs*, that Britain does, too, but you'd be wrong. We specialize in buildings that sit on the edge of towns or in suburbia or in a row in a village, which are pretending to be in the middle of nowhere. We have bred a particularly clever kind of architect in Britain, who is able, with the use of masonry, to block out the undesirable elements from our surroundings – like pylons, neighbouring garages, sewage works and housing estates – and instead focus our gaze, with the use of glass, on the pretty view, the precious thing that we bought the site for in the first place (see Howth House, pages 152–3).

There are exceptions, of course. I've filmed up Welsh mountains and on Northern Irish lochs, in remote Scottish glens and English moorland. But this is one of the most densely populated countries in the world, with a planning system creaking under the strain of new construction everywhere. In the British Isles you're never far from an A road and a bungalow – even in Wales, and especially in Ireland.

Because there is so much pressure on land, there is a reverse pressure not to release it for building, which means that, for the moment at least, the Green Belt remains sacrosanct and new projects have to happen within the planning envelopes of existing settlements. The only way to get round this is to convert an agricultural building (if you're lucky enough to get planning permission to do so)

or to swallow hard, take the expensive route and buy an existing horrible house in a beautiful place – then demolish it. That's exactly what Jon and Terri Westlake did near Peterborough (see above and pages 166–7), leaving them precious little cash to build what they wanted. Mercifully, their architects, Spacelab, designed for them the simplest structure you can possibly build: a box. A beautiful box, mind, that

previous page: **Blast Architects' house overlooking Loch Long at Kilcreggan.** above: **Spacelab's Rural Box in woodland near Peterborough.**

took the one good view, framed it in that British architecture way, and shaped it with an entire wall of glass, giving them a million-pound aspect for a budget price – an Imax movie in a theatre with just a couple of seats. This was a house with precious little inside but with a big connection with the outside, and an overhanging protective ledge to keep the worst of the sun off in summer and provide protection from the wind all year round. This is a popular idea these days: connecting inside and out, getting that 'flow', making the most of 'view'. So popular that it has become an obsessive feature of almost every new rural home I visit. No one of this generation says, 'I like half-light, dusk and cosiness and I want my house to reflect that.' They say, 'I want light, I want the place to be flooded with light, I want space and I want connection with the outside world.' We should all, clearly, be reading Wordsworth.

Two of the most romantic houses I've visited take this principle of connection to the absolute. Neither is large. One is the 'upside down' cruck-framed oak, steel and concrete house designed for Tony and Jo Moffat by Blast Architects in Glasgow (see pages 155, 168–9 and 235), which sits above the shores of Loch Long with a spectacular view out towards the Isles of Bute and Arran (and a spectacular view of Britain's nuclear submarine fleet, which variously surfaces for rearmament and supplies further up the coast). You wouldn't know that this place sits in a long terrace of otherwise very ugly houses above the village of Kilcreggan, because there isn't a view out of their house in any direction that reveals the excrescences. I say that, but being eight inches taller than the owners, I did notice one view from a side window that, bizarrely, overlooked, just, a distant housing estate. It was when I bobbed down to their height that I realized that the architect (also 6ft 3in) had specifically created this window for them and them only. All they saw, day in and day out, was a magnificent view of heather-clad mountains and sky. That's bespoke design.

The other house (opposite and page 73), almost my favourite house ever, is the ecological home built by KAP architects for Theo and Elaine Leijser, near the Campsie Fells northeast of Glasgow. This is a modest home, a simple, super-

> This is a popular idea these days; connecting inside and out, getting that 'flow', making the most of 'view'. So popular it has become an obsessive feature of almost every new rural home I visit.

facing page: **KAP's cedar-clad home, for the Leijser family near Glasgow.**

insulated, timber-framed, timber-clad box, but one punctured with deep openings that steer the eye outwards to drink in the views. The deep reveals of the windows are like picture frames, containing and controlling the view so that standing in the house is almost like standing in an art gallery with 4-D art in front of you. The gallery effect is heightened by the wash of toplight from a one-metre-wide glass panel that runs the full width of the building on the flat roof. For once, this house really is 'flooded' with light, to the extent that you begin to perceive, reflected on the walls, the changing colours of the sky.

The Leijsers' house, although another timber-clad box, fits its setting majestically. The wooden cladding is made of hand-split cedar shingles, or 'shakes', that give the building as random and as interesting a texture as the forest behind it. Making houses fit where they are, like this, is a real art. In the urban setting this has as much to do with knitting structures and materials into what is already a thickly layered and textured mix, a sort of stratified urban rock face; it's challenging and complex. In the suburbs you can argue that the process is simpler, in that houses can be allowed to perform as individual statements more freely, while also setting up a dialogue with surrounding buildings and with that peculiar form of tamed nature that typifies the suburbs, urban foxes and all. In the open countryside or on the very edges of our communities, the job is hardest of all. Not because the context is so difficult to figure out (houses like the Leijsers' use timber and glass to reflect their surroundings, metaphorically and literally, and these materials are now so common in rural architecture that we're in danger of being remembered as the Age of the Wooden Box), but because, as I've mentioned, planning permission is so difficult to obtain.

Architects often have to contextualize their buildings like hell to get planning permission for sites in open countryside. They use slate and stone, timber and grass to 'half-lose' their houses in the landscape. This isn't just expedient compliance; it also fits our current tastes for the bucolic. We're drawn to the idea of Teletubby living, as expressed in Future Systems's 222 house in Wales (see opposite and pages 170–3), an atavistic blend of primitive earthworks with glass-walled 'connectedness'. What more could the modern countryman or woman want?

Architects often have to contextualize their buildings like hell to get planning permission for sites in open countryside.

facing page: **Future Systems 100 per cent contextual house in Wales, built, apparently, from a hill and some sky.**

Mercifully, though, the supremely gifted architect can still woo the planners to let them build more challenging, assertive and downright proud and upstanding buildings in open countryside. Ken Shuttleworth, he of the Gherkin, succeeded with his own home in Wiltshire (opposite), a brilliant curving embrace of glass and concrete that positively enhances as well as literally reflects its setting. A building that doesn't embody, but actually redefines where it is.

For the main part, though, architects stick to timber and glass as their principal clothing materials for new architecture in the countryside, not grass and glass. I applaud this for functional as well as aesthetic reasons. Timber-clad buildings are often timber-constructed, too, which is a highly sustainable method of construction. Rob and Nicki Parkyn's Cornish house designed by Noel Wright (see above and pages 174–5), strikes the balance between completely hidden and part-camouflaged. It's cut into a hill to minimize its presence on the skyline – but it's still perfectly visible from some angles. The roofline follows the slope of the site. And the cladding material? Well, predictably, that's timber. This is a modest rural house, confidently embedded into where it is.

The opposite approach can be seen in Butterwell Farm (see overleaf and pages 176–7), belonging to farmer Philip War and his wife Caroline, chief executive of Linking Environment and Farming (LEAF). Designed by the architect, Charles Barclay, who describes this big barn-like wooden box as 'a prototypical farmhouse for the twenty-first century farmer, designed to be put up cheaply in a rural setting,' this place shouts assertiveness. Although, yet again, this house is timber-clad, the style and siting represent a gear-change in design. The building plays on the language of lowly agricultural and rural industrial buildings, like barns and sheds, but it also hybridizes this language with that of the twentieth-century house. It's a new strain of rural architecture, crossbred from familiar patterns but producing a style that is something entirely fresh and vigorous. It's a type of building we need to see more of in the countryside: this is New Vernacular.

above: **Noel Wright's Cornish house for Rob and Nicki Parkyn.**
facing page: **Ken Shuttleworth's own uncompromising house in Wiltshire.**
overleaf: **Butterwell Farm by Charles Barclay.**

Spacelab's Rural Box
Peterborough

This house is the modern dream of loft living translated to the countryside – real urbs in rus. This was Spacelab's very first project but it bears all the clarity and incisive genius of a mature work. It's simple, bold and inexpensive: a softwood-clad super-insulated timber box made with everyday materials but handing its owners big ideas, like a first-floor gallery which delivers first-floor views plus ground-floor full-height appreciation of the view. The doors also slide into the walls for an uninterrupted flowing connection to outdoors.

Kilcreggan
Loch Long, Argyll

Blast Architects in Glasgow work harder than most to contextualize their buildings, employing oak frames and cladding materials that poetically reflect – rather than mimic – the surroundings. So Tony and Jo Moffat's house is roofed in sheet zinc, a material that changes colour and character with the variable conditions of the sky and sea. And the vocabulary of its design is deliberately rich and ambiguous, drawing on traditional long houses, Scandinavian house types, boathouses and even upturned boats themselves.

Future Systems 222 House
Wales

It's almost impossible to imagine a building more hidden or absorbed into its landscape than this one, built on the west coast of Wales by Future Systems. It marries the mirror-like glass walls of Ken Shuttleworth's own house (page 163) with the anonymity of the Teletubbies' hill, like an ancient tumulus that's been sliced through with a glass knife.
It is, of course, absolutely uncompromising – a dazzling and arresting man-made moment in an untamed wilderness. And boy, does it have a view (overleaf).

Parkyns' House
Port Isaac, Cornwall

The Parkyn family had very specific needs for their house in Cornwall, partly to provide for their disabled daughter and partly because they wanted their house to sit very quietly in its setting. The result, designed by Noel Wright, is a building that is almost invisible from the nearby village, thanks to the way it is half-sunk into an opening in the hill and fitted with a monopitch roof that carries the line of the hill on down. And it still delivers height, light and space.

Butterwell Farm
Looe Valley, Cornwall

This is another hillside house, accessed in this case directly onto the first floor, and one that stands, by contrast to the previous project, proud in its own setting. The restrained design, simple forms and modest use of materials like wood, slate, copper and zinc all define this building as absolutely of its time. This is a hybrid of modern and traditional vernacular. This is new vernacular.

New Use

It may be the vagaries of my own taste, it may just be everybody's zeitgeist, but I'm aware that the houses I've chosen in this chapter bear a family resemblance. In New Urban (page 96) the houses seem, in the main, invisible. Obviously the best way for me to avoid having to identify new trends in urban building is by choosing houses that apparently don't exist. Or which at least don't have any façades; or if they do, the façades seem fake or are broken up, faceted and painted black, and the place is called the Stealth House (invisible to radar). New Life (page 202) features houses that are all, to some degree, still falling down, so pretty soon they'll be off everybody's radar. In New Suburban (page 130) all the houses want to be tree-huggers and move to the country. Whereas the New Rural houses I've selected, which are already in the country, all look the same, it seems – wooden boxes with big windows.

previous page: **Violin Factory, Waterloo.**

above: A true stealth house, Simon Conder's 'Vista' is a black, neoprene-clad bunker that's actually a rebuilt fishing shed. It looks like a NASA experiment ready to take off.

In contrast, in this chapter, New Use, I'm mindful that there is very little that links the projects I've chosen with each other. Aside from the fact that they're nearly all buildings that, when they were built, were never actually intended to be lived in. One was a shed, one a barn, one a schoolhouse, one a factory, one a waterworks and one a roof space dedicated to air-conditioning units. It seems that we British specialize in the eccentric adaptation of defunct and useless old buildings into homes, perhaps because good building plots are so rare in the UK.

But the credentials of converted buildings are very good. They reuse structures that often have a historic value, even if only by contributing a bit of variety and texture to a place's skyline. Once left empty, buildings like Fort Dunlop by the M6 near Birmingham (which has been salvaged and is now being converted into apartments by the developers Urban Splash) deteriorate appallingly quickly. If they're not rescued, they soon reach the point where demolition seems the only viable solution. And then, whoosh! Another piece of undervalued industrial history disappears.

Converted buildings are also usually highly sustainable. They tend to reinvigorate a surrounding area and can be environmentally efficient because the structure already contains vast amounts of embodied energy that was locked into the materials at the point of manufacture. English Heritage has calculated that for every terraced house demolished as part of the Labour government's Pathfinder scheme, the embodied energy lost in the demolition and then required to rebuild a similar-sized house on the site equates to driving around the world one and a half times.

Not that remodelling an existing building need leave any visible trace of what was there before. One creative solution is to bungalow-gobble an existing crap house, reuse its structure internally in your new building and create a brand new super-performing casing. Or you could do as the award-wining architect Simon Conder has done with Vista (see pages 180–1 and 190–1), a house near Dungeness, which began as a fisherman's shack. Instead of gobbling it, he coated it with thick black rubber, rather as a master chocolatier would *enrober* a knobbly

above: **Chris Jones and Leanne Smith's epic waterworks near Bolsover, built to emulate the utility style of Giles Gilbert-Scott.**

walnut. What was rough and poor is now a beautiful and sleekly crafted home, rebuilt around the old frame and appropriately fitted out in pale plywood, like a boat might be, or a timber-framed submarine, which is what, for all the world, it looks like.

Working on an old industrial or agricultural hulk doesn't just provide you with an interesting set of spaces but with a whole range of inspirational ideas for how to furnish, decorate and style the building, as Conder found. At the Bolsover and Chesterfield Waterworks (see pages 183 and 192–3), Chris Jones and Leanne Smith found that the epic scale of their new home dictated not only the scale of their furniture but also its language. So they built their desk out of a Mini and used bits of giant pipework, salvaged from the building's previous incarnation, as furniture around the place. This industrial chic is very cool, but only when there's a good authentic reason for it being there. And ex-utility buildings can provide the right setting and the right raw materials if ancient pumps, tanks and generating equipment have been left for you to dismantle. These buildings, although not common, come onto the market regularly, as power companies upgrade their equipment and water utilities replace local pumping stations with longer-distance, more centralized modern pumps.

But the pains of adaptation can be agonizing as you try to stretch a building this way and that. Reuben Welch and April Marr devoted months of slavish care to their nineteenth-century schoolhouse in Leith (see above and pages 194–5). They replaced and repaired stonework, faithfully copying the original mason's marks. They learnt to form leadwork to levels of craftsmanship of which any self-respecting traditional plumber would have been proud. April hand-cut thousands

above: **Reuben Welch and April Marr's salvaged and restored schoolhouse in Edinburgh.**

of reclaimed local roof slates to dozens of different diminishing sizes to reproduce the traditional graded roof that the building would have once had. They gave their souls to the shell of that building and then it thanked them by dumping them indoors without giving them a clue how to convert the place.

above: **The unassuming entrance to Milko Ostendorf and Louise McDonnell's stonking violin factory conversion in Waterloo.**

They were clearly shocked at being so rudely abandoned without the building giving them so much as a pointer as to how it might be divided and reused. Having spent a year repairing and restoring the structure, they then ummed and aahed aimlessly for months while figuring out a sensible way to lay out the interior.

Don't think that having an architect on board will necessarily help. Louise McDonnell took on so many architects, she formed her own practice, and still her violin factory in Waterloo (pages 179, 185 and 196–7) refused to come good. The intransigence of this building lay in its refusal to be dragged out of an industrial slumber. Local opposition to change and the complications of several party wall agreements slowed progress and stretched the schedule out over three years. Louise and her husband Milko Ostendorf patiently waited and carried on funding the project until it eventually righted itself into a shimmering internal space, part winter-garden and part giant hall, like a smaller domestic version of the Tate Modern but with the glamour of a luxurious open-plan hotel. This place and the Bolsover Waterworks have both rewarded their owners with an abundance of room; air around and above their owners, enveloping them in one of the great luxuries of the twenty-first century: space.

In pursuit of space, both projects have also been pushed up through the ceiling of the building onto the roof, to engage with the most open bit of air around any of us – the space above our heads. Going up is an obvious way of exploiting real estate and capturing more room, and yet, like the underground Anderson House (see pages 116–19), we don't think often enough about exploiting the underused roofs of our buildings. Lee and Laura Hughes lived in a top-floor apartment until they employed architects Project Orange to crane in a prefabricated structure on top. The resulting building is an exquisite pavilion, in its own rooftop garden, that resembles something from the Mediterranean rather than Hoxton (see opposite and pages 198–9). It may have cost them £250,000 but it has transformed their flat – and their lives – giving them views, privacy, open internal space and that connection with the natural world that is so often missing in the city. As Laura says, 'Having so much sky is fantastic.' What price can you put on that?

If one of the great devices of architecture is to borrow beauty from nature, then one of the great advantages of converting an existing building is that you can borrow from its history and allow its story to resonate through what you do to it. I was always impressed by how Chris and Leanne left the bare brickwork and flaking ceilings in their waterworks – even if they did cover up the fabulous terrazzo floor and dados. And it is impossible not to be impressed by the barn conversion wrought by the architects Hudson Featherstone in Haveringland, near Norwich (see overleaf and pages 200–1). This is an exemplary barn conversion. Not only is the building's history borrowed and then allowed to speak here, it is amplified by careful repair of the old structure using sympathetic materials, agricultural fencing and detailing in 'farmer's' galvanized fittings. New components are built in new ways and with new materials, but always in proportion and, more importantly, in deference to the old place.

This kind of clear relationship between modern architecture and old building, contemporary details and old textures and surfaces, new use and old history, feels good and fulfilling in barns that are sensitively converted like this. The relationship of then and now seems so easy, too – not at all difficult to get right, until you try it. It takes a delicate and sensitive approach and real architectural genius to do this kind of stuff. If planners were given this type of barn conversion to approve, they'd let it pass every time.

above: **A refreshed and almost surreal Hoxton skyline.**
overleaf: **Hudson Featherstone's Quaker Barns conversion at Haveringland, Norfolk.**

'Vista'
Dungeness, Kent

There are places in Britain where you could never expect a hope in hell of getting planning permission. Open wilderness, like this Dungeness beach, for example. Simon Conder's solution was to convert an existing 'cottage' (in reality a hut) into a new dwelling, still built on the old frame but radically remodelled to look as though it had been designed by aliens as a lookout. They must have felt at home. 'Vista' looks out over a moonscape.

Chris Jones and Leanne Smith 'reindustrialized' their rather polite super-size waterworks, stripping plaster back to the brick and reinstating new Crittall windows. They can now luxuriate in a vast untameable space, the next best thing to living in the Tate Modern's turbine hall.

w Use . **193**

Having slavishly conserved and
gently restored the outside of
their nineteenth-century
schoolhouse in Leith, Reuben
Welsh and April Marr were left
with no idea what to do with
the open cube of its interior. The
solution, eventually, was inspired:
a half-first floor with a curving
balcony and spiral staircase, and
an open-pyramidal ceiling at the
top. In the end, the open cube
remained more or less just that.

Violin Factory
London

Louise McDonnell and Milko Ostendorf's grand design in Waterloo became notorious for its overrun, its party-wall disputes and its £34,000 cooker that didn't work (opposite). I remember it for the brilliance with which it looked upwards to connect vertically with the great outdoors, and for the crystal perfection of some of the building's components, placed alongside the brick walls and old timbers of the former factory.

Project Orange Rooftop
London

It may seem extreme to spend £250,000 extending your flat vertically, instead of moving – but look at what you get. This apartment doesn't look as though it's in east London, or even Britain. Lee and Laura Hughes have got a slice of Manhattan or Barcelona, plus a garden, views and a 180° skyline. Cheap at the price.

Quaker Barns
Haveringland, Norfolk

Bad barn conversions abound. In fact, they're so normal that most planning departments have decided to just ban barn conversions altogether. So thank heavens for this one in Norfolk by Hudson Featherstone. Not a horse brass in sight. Spaces are left open, new build is exposed as such and light is carefully introduced with bespoke and discreetly sliced openings. It's so good, I'd like to live there.

New Life

To live in an old house is to live with an old house.
It requires a degree of unrequited love, a romantic
disposition, and a distracted and remote state of mind.
It demands on the one hand the rejection of many twenty-
first century pleasances, such as 1) draught-free warmth,
2) underfloor heating and 3) double glazing. On the other
hand, it also demands the embracing of seventeenth-
or eighteenth-centuries exigencies and the acceptance
of many twenty-first century inevitabilities, such as
1) a new roof, 2) dry rot and 3) mortgage foreclosure.
In addition to sacrifices of comfort, old houses require
love and quite a lot of money. And although they can
reveal great joys and provide great style, they do not
bend themselves easily to modern living.

Many people buy an old farm or cottage in the belief that they can drag the building into the modern age – damp-proof the walls, pour a concrete sub-floor with damp-proof membrane, reglaze and replaster it and fit underfloor heating, and bingo: get the best of all possible worlds, old and new. But what you end up with is a modern interpretation of a farmhouse or cottage, a sort of badly performing theatrical set. Flagstones, once lifted, can never be reset with the same precision, so the result is a one-centimetre grout gap like the ones in the dwarves' cottage in *Snow White*. Damp, once contained under a concrete plinth, will often rise through the wall ferociously through siphonic pressure, causing problems that can be avoided by letting a floor breathe or be ventilated. You'd be better oF building Mrs Tiggywinkle's cottage from new, in concrete, like some developers do, if that's the kind of clean, easy-to-maintain look you want.

previous page: **Castle Gwydir, Wales.**
above: **Court 15 Back-to-Backs in Birmingham.**

I for one would not reproach you for that. Because I'd sooner see historical buildings of all kinds conserved and sympathetically and gently repaired than I would ever like to see them 'restored' or 'renovated'. The very word 'restoration' implies that someone's had a go at 'imagineering' some kind of fake historical past into being: distressing the beams and fitting shiny brass handles everywhere. 'Repair', however, suggests just that: keeping as much of what is already there, of whatever period, gently repairing it and adding only where necessary. That means Tudor latches in a Tudor house, but also eighteenth-century door handles and the odd 1940s Bakelite knob, as well. Yes, all in the same house.

The reason for this approach is simple and entirely justifiable. An old building is like a book in which each period of habitation is like a chapter, each contributing to the story. Our generation is writing the current chapter: we have almost an obligation to lightly add a little of our time, in the form of modern plumbing, maybe, or in the adaptation of an old dairy into a Pilates room.

When we buy an old place we are also entering into a contract, a bargain with history, that we won't utterly rewrite or bowdlerize, or remove previous chapters and kill the story of the building dead. Of course, there are exceptions to this commitment, exceptions that are tortuously wrought out in debates between architects and conservation officers and owners of old places all the time. Often, big clunky chapters, like a badly built 1950s extension on the side of a perfect Regency villa do, and should, get pulled down and replaced with something more discreet, or with nothing at all. And then the book is re-edited and packaged as a new edition – but often with photographs of what the house did look like, just for the record. So that what it was, even if only for a brief, ugly period, is never out of print.

The big mistakes people make, and all the problems they encounter, occur when they try either to add too big a modern chapter to an ancient building, rewrite its past with some imagined and unsubstantiated history, or just rip out its guts and knock through. That's like ripping out every other page in a novel: there's a story there but it doesn't make sense any more.

Some people talk of the 'character' of old buildings, but that's rubbish. What we perceive as character is really a subtle skein woven from two to three thousand tiny details and irregularities, worn surfaces and visibly layered histories. The accrual of all that detail and those layers will have taken decades, centuries, but they can be wiped away in a moment of thoughtlessness and with them goes the 'character'. You just wield some blockwork and concrete, and there go 1,800 tiny details; fake flagstones and plasterboard will destroy another 500; PVC windows are like Domestos: they'll kill 100 per cent of known historical details dead.

Other people talk of old houses as though they have souls. That's also rubbish. What they certainly manifest is energy and human commitment. Every time I visit an ancient castle or house or even an earthwork, I'm awed in my imagination at the time and physical labour it must have taken to erect the thing, from mud and clay and rock, in an age without steam engines or electricity. The power of human beings to sculpt and manage this planet is phenomenal, whether it be embodied in Lord Foster's concrete Millau Bridge in France, or the Aswan Dam, or the Great Wall of China. And these structures are often also the result of great political will and power, built with slave labour and in difficult, dangerous conditions. Even the lowliest two-bedroom weaver's cottage in Malmesbury represents weeks of toil and craftsmanship, from collecting the stones from the fields, to shaping door lintels, to digging clay for tiles, to carving window frames, to forging door handles. Could you do that? I couldn't.

Every ounce of a man-made object – and buildings are pretty big man-made objects – has a value. It represents a human being's time and devotion. A fact that I find pretty humbling, not least because for 15 years I was a designer and maker of furniture and lighting. I know the effort that goes into making a chair or even a spoon, the weeks of time spent getting something right and then weeks of time working with other people to make sure they get it right. So I try never to take for granted the human energy that is embodied in everyday objects, houses included.

If buildings reverberate to anything, it is to the energy and commitment inherent in the craftsmanship of all the details in the place. A Georgian skirting board will have started life being sawn from the forest by hand, then perhaps roughly adzed into shape. It'll have been planed by hand, then pencil marked and cut, with a number of small specialist hand planes, to the finished profile. Oh, and then sanded by hand. Compare that series of slow, hand processes with those used to produce a modern standard torus skirt. The wood will have been cut, sawn, planed and profiled all by machine. Human effort required in 1720 to produce one metre of skirting board = 10 hours. Machine effort required in 2006 = 3 minutes. So what the hell are we doing when we rip out an old moulding and replace it with a new one? We're devaluing the building, devaluing the human effort that went into it in the first place, and actually eating into the thing that some call 'soul', the *genius loci*.

That *genius* is a fragile thing. Gone with the skirting board is a paragraph from a chapter in the house's story. Gone is the craftsmanship and human effort embodied in it. And gone, too, is the regional flavour of the work. In describing how communities and architecture can be made to feel rooted and part of a place in 'A Plot on the Landscape' (see page 64), I explained how important the changes in landscape and geology are in determining unique and local characteristics in buildings, and that vernacular buildings vary wildly across Britain, often in the space of just a few miles. The *genius loci* of places are bound up in the thousands of physical details that make up that place. The special character of old buildings is also hugely dependent on local physical conditions. We love old houses for their diversity, which is born partly out of the history of their use and partly out of the landscape they're in: the room sizes, window details and render finish of a cob and thatch farmhouse in Devon will depend, fundamentally, on the engineering properties of those building materials, which will have been sourced locally. Those properties will be very different from those of a Somerset farmhouse 40 miles away, built from stone and tile, and as a result the houses will feel vastly different. *Genius loci* is in the mud.

So, the specialness of old buildings is partly to do with 'historical detail and style' and the building's value as a document of one time, or several times, in its history. It's partly enriched by the storybook quality, in the layering of history. It's also partly made unique by the locality, local materials and by the limited geographical spread of a particular building style. And it's partly the human energy embodied in the building's craftsmanship and details.

We love old houses for their diversity, which is born partly out of the history of their use and partly out of the landscape they're in.

Of course, none of these are objective values. They all need our passion and enthusiasm for history and for the built environment to wake them up and turn into living ideas. I have no problem with that because I studied the history of buildings at university and live in a 500-year-old house. Like any conservationist, what underwrites my enthusiasm isn't academic interest but romance. The reason I love old buildings is because I can imagine so much when I'm in them. And it's that romantic disposition that fuels all the above reasons for getting excited about old houses.

But to whom can you turn to back up passion and enthusiasm with help and guidance? Your mortgage company? No. They'll want to see a concrete sub-floor and damp-proof membrane everywhere. A local surveyor? Well, when we bought the farmhouse we now live in, I got a local surveyor round who simply suggested we gut the place back to the four walls and start again. We didn't, and ten years on the gnarled and sagging oak beams, old plasterwork and perforated elm beam upstairs are all still there. With a fair wind they'll still be there in 100 years' time.

In truth, there are very few people to turn to. Builders will always want to remove a window and replace it rather than try to fettle and repair it. There are other dark arts, too: that of laying a lime-ash floor, constructing a cob wall, or thatching with straw rather than reed. Finding a good builder is hard enough, but the chances of finding specialists to perform these kinds of miracles sometimes seem as likely as finding a hoard of medieval gold behind the wainscot panelling.

Unless you're prepared to join the provisional wing of the conservation movement, the organization which, in effect, has underwritten conservation philosophy since its inception in Britain, and which, in a world of changing values, 'heritage' industries and fudged government policies, is the guardian of the eternal flame of truth. The SPAB (Society for the Protection of Ancient Buildings) was founded by William Morris in 1877, partly in response to the overzealous stripping back and remodelling of medieval churches by Victorian Gothicists. Morris's fiery manifesto for the Society is as valuable now as it was then, advocating the careful and considered repair of buildings rather than their wholesale 'restoration'; suggesting that repairs should be both sensitive to the character of the building and also obvious; and maintaining an almost archaeological reverence for the way

facing page: **Gentle repair with sympathetic materials like oak, stone, lime and distemper lies at the core of conservation philosophy.**

in which time and successive generations have made their mark on architecture. Of course, his view was fuelled by romanticism, just as any building historian's is. But it was also fuelled by indignation and by a rigorous and almost academic appreciation of the historical and documentary value of historical architecture: not just as, say, an example of 'Perpendicular', but of how the changes in society are borne out in the alterations and changes of use in a building.

Even today the Society often finds itself embroiled in complicated arguments over whether to save, demolish, adapt or remodel an eighteenth-century addition to a medieval building, or how a brand-new modern addition to an old structure should be designed and integrated. The reason it gets so embroiled is that it's an amenity body: it has to be consulted as part of the Listed Building Consent process. If you apply to do something contentious to a Victorian House, then the Victorian Society or 'VicSoc' will get involved. There are other bodies like the Georgian Group and the Twentieth Century Society, each as valuable and each with their special period, and, of course, the heritage police in the form of English Heritage and Historic Scotland. But they all rely on the SPAB for their philosophical inspiration. The SPAB is, bar its equivalent body in Norway of all places, the oldest conservation Society in the world.

I particularly like the Society, not just as a champion for the sympathetic repair of historic buildings, but also because, as a result of its philosophical purism, it has emerged in recent years as a champion for modern architecture. Odd, I know, but crudely put, the SPAB often finds itself in a position where it would rather see a contemporary glass, oak and steel restaurant/foyer/visitor centre sensitively placed on an ancient site than a pastiche building that mimics the earlier architecture. There's something refreshingly clear about this approach, a rigour that encompasses not just the modern view of our history but also the future's view of our time as part of history.

But all this is background. The point I am slowly getting around to is that of all the amenity bodies, the SPAB offers by far the greatest number of hands-on

> The SPAB often finds itself in a position where it would rather see a contemporary glass, oak and steel restaurant/foyer/visitor centre sensitively placed on an ancient site than a pastiche building.

courses for homeowners and professionals alike, from 'lime days' to woodworking workshops, and it publishes a host of technical manuals on the care and repair of old houses. I can't recommend them highly enough.

As an owner of a listed house your first encounter with the 'heritage movement' will be in the form of your local authority conservation officer. That's if you're lucky enough to have one, because one in three councils doesn't (they're not a statutory obligation). Conservation officers are generally both immensely well trained and also underpaid, two qualifications that in the worst case can lead them to be bitter and twisted individuals, constantly resentful that people with more money than sensitivity (and more money than them) are buying up and destroying the built heritage around them. Mercifully, most are passionate and considerate people who, if approached in the right way, can become your ally and source of help as you plough that lonely furrow of historic homeownership. Don't see them as the enemy, but learn to speak their language and make them your friend. If you have no conservation officer then more than likely you'll be dealing with a junior planning officer who, given his very presence, will imply a local planning department that is underfunded, understaffed and overstressed. My only advice in these circumstances is to go on a SPAB course, and perhaps pay for the hapless junior to go on it, too.

What follows is a series of buildings – and interiors – from different ages, which between them exemplify some of the ideas that I've written about here. Buildings that have been repaired sympathetically, sometimes rebuilt and very occasionally even restored in the spirit of dramatic reconstruction. But always with an eye for what constitutes good conservation. And always with sensitivity for the way in which the real charm and character of our old houses reside in the delicate layering of detail, texture and material.

One I particularly like is the Court 15 Back-to-Backs project completed by the National Trust and Birmingham Conservation Trust in 2004 (see overleaf). For once, this is the National Trust producing a modest heritage project that has some life and vitality about it. Or maybe that's the Birmingham Conservation Trust, which pretty much drove the project all along. In any case, this is a rare collection of dwellings, painstakingly conserved and each furnished from a

different period in order to better understand the history of these modest gated communities of the nineteenth century. What's amusing – and appropriate, given that the houses were occupied for decades – is that some details have been left as they are, some repaired, some 'restored' (for example, the occasional newly made reproduction window) and some hypothetically installed as pieces of theatre.

This is an approach being taken at Skipton Peel (see opposite), a project that I've included here despite it not being finished. The owners, Francis and Karen Shaw, are passionate castle fetishists and they have poured their souls into this project with support from English Heritage. The place was a roofless ruin, but where they could they've repaired stonework and replaced it only when necessary; in other cases they've had to restore Georgian timber windows following engravings. Elsewhere, where all they have is a pile of stones, they've had to imagine rooms by visiting other buildings of the same type (there are almost none). I'm looking forward to the finished result. Whether I like it is almost

above: **A restored interior of Court 15 Back-to-Backs in Birmingham.**

irrelevant: what matters here is the rigour and flexibility of their approach and being, unusually, allowed to rebuild a Scheduled Ancient Monument. This project may well be a test-bed for a move within English Heritage to see more derelict ancient ruins rebuilt, rather than simply stabilized and 'conserved'. After all, buildings were put up to be used.

This flexible approach was used in the rebuilding and repair of Sker House, another epic conservation project, by the Buildings at Risk Trust, of one of the few great Tudor houses in Wales (see overleaf and pages 220–1). More was left of this place than at Skipton Peel, so both conservation and replacement processes followed more conventional lines. The result is a beautifully crafted but simple reinstatement of what was there, a towering edifice of stone, slate, wood, iron and lime. This is the kind of project where conservation methods really pay off: because so much is retained and repaired, you don't feel as though you're sitting in a reproduction of an ancient building but in one that still resonates to its own history. And because sympathetic materials like lime plaster are employed, the character of the place is less hard-edged and more 'waney-edged': softer and bumpier.

Of course, if you leave it alone and don't do anything to a house, it's going to look extremely waney-edged, like the ceramicist Rupert Spira's farmhouse (see pages 216 and 218–19). Short of the roof caving in and an elder tree sprouting out

above: **Skipton Peel before work started.**
overleaf: **Sker House in a similar condition.**

of your bed, this is about as untouched as historic houses get, at least those that are still inhabited and where social services haven't been round. Plaster walls have just been gently relieved of only the largest flaking chunks. Wallpaper has been delicately glued back here and there. Repairs, whether in lime, Polyfilla, glazing putty or chewed-up bits of the *Daily Echo*, are left visible. It's more than a little primitive, difficult to look after and actually in need of some more attention; and I love this place, not least because my own ancient house resembles it in more than one way. There's a romance and studied modesty about this approach and a relaxed, open attitude about what matters. I wish more old houses were left 'not quite finished' like this one, because in truth they never are finished.

For that reason alone it would be worth including Castle Gwydir here. Gwydir (see opposite and pages 222–3) sits in what I think can safely be referred to as deepest Wales – deep as in nigh-geographically unreachable and also as in slightly hidden and lost in time. This is a house to which each generation has done something, making it a palimpsest. It feels like a historical document, but not one written by a desiccated academic and printed by Yale University Press in very tiny type with a grey paper dustsheet. This is precious. It's a vellum parchment, illuminated in tempera, much overwritten, scratched on, rubbed away, part destroyed and then part glued back.

above: **Rupert Spira's farm.**
facing page: **The courtyard of Gwydir Castle.**

You can visit Gwydir and hire it for weddings and Bar Mitzvahs. It's worth getting married – and getting married in darkest Wales at that – just to experience the magic of the place. Peter and Judy Welford, the owners, were determined not to show Gwydir as some kind of bared-stone, suits-of-armour Pot Noodle version of what a castle should be (a previous owner had tried that in the 1950s), but as a historical home with decoration and comfortable furnishings. Yet they have also been delicate in handling the intense atmosphere and power of the place, not wanting to disturb elements of the building unless absolutely necessary. They were informed by Peter's experiences as a building conservationist and fed, for the years it took them, by their shared romanticism, meagre incomes and fresh air. Their dedication and the resulting miracle of a house is a reminder that in owning and rebuilding an old house, it isn't enough to be passionate about history, or to be brilliantly knowledgeable – though both of which help hugely. Nope, the prime requirement is to be a hopeless romantic.

Rupert Spira House
Bishop's Castle, Shropshire

The most extreme form of conservation is embodied in the 'Leave It Alone' movement. Rupert Spira's farmhouse epitomizes a slightly more caring approach, but one that is loathe to disturb the sedimentary layers of the building's history. Older repairs and changes have been left as they are and new work is done with sympathetic materials like limewash. The result is a journey through time that makes all the stops.

Sker House
Porthcawl, Wales

Sker House is now a home. It has a roof and glass in the windows. It is one of the great architectural rescue operations of the last ten years, despite the repair and reconstruction works suffering all kinds of setbacks. The strategy was simple: save what you can, repair as much as possible and anywhere else replace like for like. The result, although in places a major rebuild, has all the integrity that the building's saviours, the Buildings at Risk Trust, could hope for. An integrity which provides a contrasting but consistent backdrop for an entirely contemporary furnishing scheme.

Castle Gwydir
St Grwst, Wales

It is romance that drives any owner of an old house to want to live in it. And Gwydir has romance in spades. Built over several centuries, its collection of rooms, solars, chambers and halls have been gently teased from decay and slow death: original panelling was found in America and then brought back; walls and roofs have been repaired and rebuilt; and the house ghost left alone to get on with her moaning and clanking.

3 Doing

Plotting & Scheming

Without a doubt the biggest problem facing anyone wanting to build in the UK is finding a plot. There are tens of thousands of people – you're probably one of them and I know I am – who would like one day to build a home of the highest and most exciting architectural order, but who simply cannot find anywhere to do it. Even those blessed individuals who do have a meagre site usually find planning obstacles in their way that are directly related to the rarity of empty space in Britain.

We are, apparently, the seventh most populous of the significant countries of the world (you have to exclude Monaco and the Vatican City from statistics like these), the second in Europe after Holland. So our building land is rare, overpriced in a pressurized market and subject to the most vicious and lengthy processes of public scrutiny. Half of the cost of the average house in Britain is the cost of the land. In the Southeast it's more like two-thirds. In effect, we pay a land levy of between 50 and 60 per cent just for the pleasure of being able to live in our own country. Our population density goes a long way towards explaining our national obsessions: banning wind farms, restricting the growth of our towns and cities, and football, which also involves cramming large numbers of people into a tiny environment.

But being a resourceful nation, we devise ever more ingenious ruses for finding somewhere to build. Some people are lucky enough to get permission to build in their own back garden, or on their roof, or to convert an outbuilding. Others buy up old utility buildings like water-pumping stations or electricity substations to convert. There are the M-House and The Retreat, two small off-plan dwellings that you can buy and install in your garden or on a static caravan site because, effectively, they're just mobile homes (but brilliant, beautiful ones and you can't see the wheels).

One extreme tactic, unknown anywhere else in the world, is bungalow-gobbling: the art of buying a bungalow and enclosing it inside another building. Some hapless self-builders even go to the extent of buying a perfectly decent little house on a nice plot, only to demolish it in order to build what they want. That's a very expensive route.

Having tried (in an entirely dilettante way) to navigate my way around plot-finding websites like www.plotfinder, I can say with some authority that they're for the truly committed. A derelict end-of-terrace pink-rendered two-up two-down in Swansea isn't difficult to find; a virgin plot with views across the Berkshire Downs is. And to find anything approaching a gem of a site involves spending every evening trawling through these websites' databases. But who says you're not truly committed?

So arm yourself with knowledge and get everything set up so that you are in a position to buy a site as soon as the opportunity arises. In a competitive market speed and preparedness can easily make the difference between securing an ideal site and missing out. Start by talking to lenders, so that you know where the finance might come from (obtaining agreement in principle, if you can), and ask for a letter confirming this. Then figure out what funds you'll use for the deposit for the purchase. And if you don't already have a solicitor, speak to several and choose one you feel comfortable with.

Depending on your circumstances, you might need to put your existing home on the market to raise funds. That's good, because the best possible position is to be a cash buyer. In the bear market of plot finding you are likely to be up against builders and developers who will not have properties to sell in order to fund the purchase of a new property or site, and you may well be up against other private buyers, as well. Bear in mind that both vendors and agents really like potential purchasers who are in a position where they can move quickly.

Estate agents

Most land in Britain and most buildings are sold by estate agents so, regardless of your opinion of agents, you're probably going to have to deal with them. The role and position of agents is widely misunderstood and so the better your understanding, the more you can work the system to your advantage. In normal sale transactions, estate agents work for the seller and owe potential purchasers absolutely nothing. This means that it's you who has to do the chasing – it's no good

previous page: **John Cadney and Marney Moon's self-build Finnish log cabin in Kent.**
above: **Alex Michaelis's radical sunken house in London.**

calling an estate agent once to register your requirements and then sitting back and waiting for the phone to ring or for tasty sites to drop through your letter box. Agents have no shortage of individuals and builders/developers who want to buy sites, and you're probably going to be one of maybe hundreds of people on their books looking for property or land. Some agents won't even add your name to their mailing lists.

Agents, quite naturally, want to earn their commissions as quickly as possible, with the greatest certainty possible, and by doing as little work as possible. Moreover, they're interested in repeat business – in other words, where the next commission is coming from – so they really like selling to speculative developers in the hope that they can pick up the job of selling the developer's new houses when they're built.

Which agents you choose to contact depends on the type of site you are looking for. Generally, few agents regularly handle development sites: most just sell houses. Those that do tend to be the larger firms that offer a range of services, such as commercial property letting and sales, management and valuations. Any local agent should be able to point you towards these firms, and if you're looking in a rural area, get in touch with agricultural agents, as they sell barns and other rural buildings for conversion, as well as village and rural plots.

You might be considering demolition and replacement of an existing house, or bungalow-gobbling, especially if you are looking in a high-density or very popular area – in which case it is, of course, worth registering with conventional residential agents.

Shopping

Having figured out your plan of attack, it's time to start making some appointments. Ideally, make sure that you meet up with a manager or senior member of staff, one who has experience of the area in question and the market. And, very importantly, take them a copy of a written specification: a document that lists all of your needs, in terms of plot size, location, proximity to services and village or town centres, budget and anything else that you feel is crucial to the way you want to inhabit that site and live. Make sure that all your contact numbers and addresses are included and details of your ability to move swiftly in purchase, if possible. You need to distinguish yourself from the other people also looking for development sites and demonstrate that you are a serious buyer – and they will thank you for it. If you can include information about funding sources and the name of your solicitor, it will all add credibility. Whereas vagueness about where, when, what and how you plan to do anything has the opposite effect. Of course, being a cash buyer is ideal but a ready source of finance indicates that you can act quickly and are organized, all of which encourages confidence to flourish. To all intents and purposes you'll come over as efficient and business-like as a property developer.

Once you've made contact, stay in touch as regularly as possible without becoming a nuisance. Do this by phone, email and, from time to time, in person. This keeps you in the forefront of their minds and means that as soon as sites come onto their books, there's a chance they'll let you know about them. Agents all have websites where you can view properties. These are a convenient way to monitor the market and see what's coming up for sale, but emails and websites can never replace the personal touch – and the best sites are sometimes gone even before they reach the website.

A few agents stage auctions and occasionally a development site will come up for sale in this way. If you are vigilant, your regular contact with the estate agents should alert you to the inclusion of a potential property in a forthcoming auction catalogue. In any case, auction catalogues are usually put onto agents' websites, so you can see the properties on offer there. Take advice before bidding at an auction, however, because a successful bid is irreversible and demands the immediate payment of a deposit. It is also crucial that you fully investigate the site and conduct a search through your solicitor beforehand.

Browsing for a site

1. THE PRESS

Apart from the estate agents' adverts in your local newspaper, many of the the nationals, such as the Saturday *Daily Telegraph*, the *Sunday Times* and *The Times*, all have extensive property sections. These are useful for background research but they also contain adverts for sites. *Daltons Weekly* and *Exchange & Mart* contain ads for all kinds of property – from old shops to derelict wasteground as well as houses. Magazines, such as *Homebuilding & Renovating*, *Build It* and *SelfBuild & Design*, also have pages of sites for sale. These are mostly on agents' books and the chances are that they'll already have been sold by the time you see them. But they do at least indicate which agents to contact and give an idea of prices and availability.

2. LANDOWNERS

Many landowners are public and private bodies with their own estates departments that look after the organization's property and, from time to time, sell off surplus. These bodies include councils, utility companies (gas, water, electric, telecom) and commercial companies, like breweries. The MOD is likely to be disposing of land in the next few years, as, perhaps, are the National Health Service and National Rail. The properties they dispose of can sometimes be eminently suitable development sites, such as former schools, pubs, shops, car parks or just parcels of land. Some estates departments use estate agents, others market sites themselves and usually, but not always, these organizations will have already got planning permission to maximize the land value before selling. Look out for their adverts and 'for sale' boards and look up the companies' websites or phone numbers. Get in contact with their estates departments to see what's currently on their books and what might be coming up. Even if they're planning to use an agent to sell a site, at least this way you stand a better chance of knowing about it in advance. Forewarned is forearmed.

3. LAND AGENCIES/FINDERS

There are quite a few companies that cater specifically for people who are looking for sites on which to create their own homes (for example, Plot Browser, Plot Finder and Plot Search). They are usually web-based databases of sites that are already on the market with agents, compiled by reference to counties or council areas. In return for a subscription you get access to the database, but because the information is second-hand, it can be

out of date. At the very least, these websites are useful as a research tool: indicating where sites come up, who deals with sites in your chosen area and the level of asking prices.

More personalized land-finding services are dotted around the country, usually operating on a commission basis. The company or individual searches for you, relying on a detailed brief that you provide, and in the event that you buy a site they've introduced, you pay a fee, often a percentage of the purchase price. You can even ask local estate agents whether they'll offer this service for you – good agents know about other sites on the market as well as ones they're handling.

4. NETWORKING

Since there's a limit on how much time you have and how many places you can look at, it makes sense to enlist others to help in your search for a site. So tell everyone you know or come into contact with that you want to buy a site and outline what it is you are looking for. And that means family, friends, colleagues and acquaintances. For that matter, don't be proud – put up a 'wanted' poster anywhere with a notice board.

Anybody you talk to will probably mention your interest in buying a plot in conversation with somebody at sometime. In this way your intelligence network will spread out and your efforts will multiply many, many times. To increase the effectiveness even further, offer a cash reward for anyone who brings you a site that you end up purchasing. A thousand pounds added to your budget won't break your project but could make the difference between creating your own home or not.

5. SCOURING THE PLANNING LISTS

The biggest problem you face, and it's a perennial problem, is being too late. Sites often come onto the market and sell instantly. One way to avoid this and to pre-empt other interested parties is to find out about sites early – before they even reach the market. Every development site will at some point pass through the council's planning department – an underexploited but obvious place to find out about property likely to be sold. Councils publish weekly lists of planning applications submitted. These lists are often on the councils' websites but, failing that, they are always available in parish/town/community councils' offices and in local libraries. The lists normally contain a short description of each application together with the address, and you can usually divine from these scant details whether an owner of a site without planning permission is attempting to get permission and thus increase the value of the site before putting it on the market.

If there are any promising-looking applications, note the application number and address and look up the application itself. Good council websites enable you to see planning application forms and drawings online; otherwise it's a case of having to pop into the council's offices in person to study the documents there. You can usually figure out from the drawings and photographs in the application whether the site could be right for you. If so, make a note of the name and address of the site owner and the person (called the agent) who made the application on behalf of the owner or occupant. If there is an agent, contact that person or firm and ask about the owner's intentions. It might be that the owners are planning to build or convert for themselves with no intention of selling the plot. But there's always a chance that they'll be thinking of placing it on the market. And again, even if they intend to market it through an estate agent, at least you'll get first bite of the cherry by knowing about it in advance. More rewardingly, a good early offer might persuade the owner not to put it on the market and incur estate agent's fees.

Similarly, councils keep records of all planning permissions that have been granted. So you should also look through the recent permissions and see whether any of those might be appropriate, then look up the application and the owner's or agent's contact details in the same way.

Creating new plots from fresh air

Instead of waiting for sites to come on the market, you could take a much more active approach and try to identify development opportunities for yourself. You, too, can become a fully-fledged property developer. Obviously, what you search for depends on the type of project you ultimately want to carry out. If you're planning a new build, for example, you could start by looking for existing houses with large gardens (at the back, the side or in front of the existing house), orchards, vegetable gardens or allotments, blocks of garages or outbuildings. You could also research unused commercial buildings. Or disused utility buildings. In built-up areas and within permitted development envelopes, you'll find that planning permission is more likely to be forthcoming. If you prefer a rural setting, then a hunt for some existing substandard houses in the countryside is most likely to get over potential planning objections.

If you aim to convert, any existing commercial or industrial building in a built-up area has potential, provided that the council will entertain a 'change of use' – that is, into a residential dwelling. A wide variety of structures get ingeniously converted, and many of them make it onto *Grand Designs*: water towers, windmills, railway stations, pubs, shops, public WCs, warehouses, churches and chapels, Martello towers. You are, of course, not limited to empty buildings, but unoccupied property is more likely to be redundant and owners more likely to sell. In the countryside, however, conversion policies are more restrictive. It's often not possible to change the use of a place from commercial, agricultural, recreational or industrial use to residential, especially if the site falls outside the planning envelope of a village or town.

The secret to searching for new sites is to cultivate the ability to visualize what could be accommodated, rather than just seeing what exists now. An overgrown piece of scrubland looks very different when it has been cleared and cleaned up.

In order to find somewhere to build, spend as much time as you can in the area you're interested in. And don't just drive around: driving is not a good way to spot opportunities, not least because you should be concentrating on the road. Instead, get on your bike. It's no exaggeration to say that a bicycle could be one of the best site-finding tools that you employ, being a compromise between covering the ground and having enough time to study the surroundings properly. The bonus of searching like this is that you might also come across sites on the market that your local estate agents hadn't turned up.

But a bicycle isn't a magic wand. The most thorough exploration by bike won't reveal all possible opportunities, either because areas of land aren't always visible from public highways or because some sites are so overgrown that you just miss them. So, secret weapon number two is the Ordnance Survey map. Studying these helps you to see where there is space and usable gaps between buildings. Maps at 1:1,250 or 1:2,500 scale allow you to see individual buildings and boundaries in sufficient detail to figure out whether there's room for the house and plot you want. You can, of course, buy Ordnance Survey maps, but in quantity they're quite expensive. So visit the local library, which keeps copies of all the local maps – as do councils, where detailed maps should also be available for public inspection.

Finding a likely site is just the first step. Next, you need to find the owner. In some cases this is obvious – for example, where the site is in the grounds of a house. You can look up the name of the occupier in the electoral register or with the Land Registry. However, it's sometimes not obvious which building a site is attached to, and when a property is empty it can be more difficult to locate the owner. If property is registered, you can find the name and address of the registered owner from the Land Registry website (landregisteronline.gov.uk) or by phone (0870 010 8318). If that fails, detective work is needed. Ask around in the area or check historical planning records at the council and take the name of the application documents. With a contact, you can approach the owner to find out if he or she is interested in selling. A short letter just asking that question is probably best, enclosing

a stamped addressed envelope or undertaking to phone in a few days. How you handle the discussion with the owner is a matter of personal preference. You could try buying without declaring your intention and take a chance on getting planning permission, but that is a high-risk strategy and one they may see through. Alternatively, you could suggest an option or a conditional contract. An option agreement secures the site for you for a fixed period, allowing you time to apply for planning permission – but you are not obliged to buy if you choose not to. A conditional contract is a binding contract to purchase, provided that a specified condition is satisfied – in this case, the condition being a successful planning application.

Of course, a proportion of owners will not be interested in selling. A proportion will get permission themselves and/or put the property on the market (if that happens, you should make sure that you're in a position of confidentiality to know about the sale early and be ready to make an offer). And a proportion will be prepared to do a deal. Whether you score is pot luck, but if you work hard enough and make enough approaches, statistically the odds are that you'll succeed in the end.

facing page: **The view from Kilcreggan, Loch Long**

Planning Policy
& the Planning Police

Every council – or local authority – depends on a raft of government policy, guidelines and centralized prescriptive codes, all emanating from Westminster. The 'Development Control Guide' runs to 17 volumes, so it's hardly surprising that many senior planning officers are frustrated by the rigidity of a system that has had the creativity squeezed out of it. All in all, we'd be better off with policies that were regionally or even locally formed to suit the character and development history of an area; it's certainly the case that as a discipline, planning needs to find its creative core again. And as I wrote in 'A Plan on the Map' (see page 84), regardless of the internal struggles and changes that planning faces, there is no doubt that as a nation we have pretty well lost our faith in the planning process. That is, until our own application happens to be successful.

It's important to remember that planning decisions are nearly always contextual: that is, the officers, exercising the maximum of creativity allowed them, will apply policy but always in relation to the site, the location, the street, the surrounding development and their own master development plan. Every authority has a local plan: in metropolitan areas and throughout Wales they're called unitary development plans (UDPs), but the system in the UK is changing and new frameworks will shortly be introduced (doing the same job as local plans), called local development frameworks (England) or local development plans (Scotland and Wales). You should use the local plan, viewable at your council offices or on their website, as your starting point when working up a planning application.

For you as a self-builder the most important planning policies are those that specify where you can and can't build a house. In most cases, new houses are allowed inside the council-defined boundaries, or envelopes, of all or some towns and villages, as shown on the maps in the local plan. Outside towns and villages new houses are normally permitted only as replacements of existing houses, as conversions of existing buildings or, in some council areas, as infilling of small gaps between established groups of houses. If a site doesn't fall within an area or category in which the council usually allows new building, frankly you're unlikely to get permission.

Some other planning policies set out standards for design and layout. These deal with things like the quality of the design, the relationship with the setting and adjoining properties, density, safe access and privacy of neighbouring properties. Certain areas and buildings are considered especially sensitive – green belts, areas of outstanding natural beauty/ national scenic areas, national parks, conservation areas, listed buildings. These are subject to local plan policies imposing even greater restrictions and requiring designs sympathetic to the characteristics of the location. Planning policies are written in jargon so, as an alternative to looking them up, you can ask a planning officer about the policies that apply to your site. Most planning policies are open to interpretation and require subjective judgement to be exercised; they do not comprise a rigid set of rules. If your proposal complies with the policies, you stand a good chance of getting permission.

facing page: **The library tower and 'library lawn' of Jeremy Till and Sarah Wigglesworth's London house.**

The national government provides a detailed framework for councils to draw up planning policies and decide applications. This is available on the respective government websites, such as DEFRA (Department for Environment, Food and Rural Affairs) and the DCLG (Department for Communities and Local Government). But the national government also publishes 'guidance' as well as policy. Some of this guidance relates to house building, conservation and rural conversions. Good design is a constant theme of government guidance, though what constitutes good design isn't always detailed or prescribed. One useful destination is CABE (Commission for Architecture and the Built Environment), which publishes an excellent series of documents, written in plain English, setting out all kinds of design guidelines for local authorities. Councils will often have supplementary guidelines as well, which can give more detailed guidance on design and layout, either throughout the whole council area or in specific parts, for example, in conservation areas. In order to get permission, your scheme will have to comply with this council guidance as well. And if you really want to get ahead, download or order some of the design guideline documents from CABE. Some villages, national parks authorities and many towns also produce design statements for local homeowners.

What happens when you apply

Planning departments are supposed to deal with applications within eight weeks and, in fact, there is increasing pressure to do so because the eight-week target date is now the prime measure of their

performance – upon which their funding is based. Sadly, this often leads to many applications being refused simply on technicalities or on the invocation of some obscure development control regulation or guideline simply because refusal involves the least amount of administration. The last thing you want is your painstakingly assembled planning application – and your beautiful design – being rubber-stamped for refusal by a junior planning officer who's under pressure to meet targets.

When you make a planning application, the council publishes your application and then puts the application out to a whole host of consulting bodies. The opinions of the consultees and local people then influence the decision on your application. Consultees can include the parish/town/community council, the highway authority (which comments on access and transport aspects), the environment agency, the council conservation or urban design officer, and bodies concerned with design and conservation, such as Cadw, English Heritage, SPAB (Society for Protection of Ancient Buildings) and Historic Scotland. If a consultee objects strongly to your proposal, then it's unlikely to be permitted by the council.

Many consultees publish their own guidance or standards and if your application complies, the consultee would, of course, be less likely to object. And it's usually possible to speak to consultees in advance to find out how they might react to a scheme. You might even be able to work with them to overcome any concerns.

The strength of local opposition or support from residents can also be a deciding factor. Anyone can write to or email the council about a planning application. Planning officers' recommendations on applications are frequently overturned because of public pressure, as councillors are anxious to be seen to be reflecting local opinion. Prior discussion and negotiation with the parish council and neighbours can sometimes head off objections and even garner backing for an application.

Starting point

For a planning application to be accepted, the design has to fit the site, the setting and the area. The site and the building have to be in proportion, leaving a comfortable amount of space around the structure so that it doesn't look cramped (though urban plots in high-density areas may demand filling the site to its brim). There may need to be room for gardens, vehicle access, turning and parking, and maintenance access. Your design needs to be compatible with the location and reflect its context, which isn't to say that it has to be the same as the other buildings around it: some of the most interesting architecture happens when a building strikes up a countering position to its surrounding buildings – but in a conversational and complementary, rather than combative, way.

The design also has to respond to the topography of the site and the area, taking account of the shape and slope of the land, of natural and man-made features on or near the site and of the way in which the topography is altered by surrounding architecture and its 'massing'. In architecturally mixed areas applicants generally have a freer hand but in, say, specially designated landscape or conservation areas, the design parameters are more restrictive. Design, of course,

is highly subjective and the council's views can be unpredictable and inevitably based on an individual officer's personal opinions. So early discussion with a planning officer is crucial. You need to make them your ally and be prepared to learn from them.

Every component, every feature of your design is going to be a discussion point. For example, if you need to make a new point of entry onto a road, the highway authority will want to make sure that it's safe for both pedestrians and vehicles on the road: visibility is important and there are standard distances required, based on traffic speeds. Remember, too, that a new house won't be allowed to detract from the privacy of existing houses. This means that windows can't overlook the side or rear living or bedroom windows of neighbouring houses or their private gardens close to the house. A new building will also not be allowed to block light or outlook from the windows of principal rooms in adjoining properties (though outlook isn't the same thing as a particular view, to which neighbours have no right). The standards that are applied are, again, governed by the nature of the locality: in heavily built-up locations, for example, where houses already have relatively little privacy, the requirements will be less stringent than, say, in the suburbs.

Drawing up an application

Your first big decision here is who is going to help you to do this? Unless you are a designer, you'd be foolhardy to try yourself, so you'll need an architect. Depending on the complexity of the scheme, you might also need a planning consultant, a structural engineer and a landscape architect at this early stage. Decide how hands-on you want to be and make sure that everybody in your professional team is aware of this. You don't want any tiffs right at the start of your beautiful relationship. And follow the advice on drawing up a brief on pages 34–5.

A planning application can be made either as a full application (with all supporting detail drawings: thorough but expensive if you need to keep revising your applications); as an outline application (with a minimum set of drawings and few details); or as a detail application (following the granting of outline permission). Outline permission is always best if there's any uncertainty as to whether you'll get permission or if you want to test the principle without needing to have a design drawn up in detail (though outline applications can't be made for conversions).

Where there are reasonable prospects, make a full application. There'll be several forms and certificates to fill in, all of which can be obtained from your local planning department. You need to start the design work by commissioning an accurate site survey that has been measured with levels. This will provide a sound basis of information for the design and the necessary drawings. A full application involves submitting drawings showing the position of the building within the site, its elevations (sides) and its floor plans. Sometimes sections through the site and building are useful additions. For less conventional designs, models often convey the concept of a building better than drawings, especially to those not used to reading them, such as neighbours and councillors (who will be the people sitting on your local planning committee). It isn't good enough to submit a few technical-looking lines on a piece of paper: good presentation of schemes is vital, in the form not just of an attractive and realistic-looking set of drawings with a bit of flair about them, but also a professional well-written supporting case. Your written statement case should comprise a design statement, setting out the factors taken into account in formulating the design and explaining the design concept, a sustainability statement and a justification by reference to planning policy, supplementary council guidance, government guidance and site-specific points. The more you can quote, and comply with, the kind of documents that planners refer to – such as those published by the DCLG, amenity bodies such as the SPAB, and Britain's design authority the CABE – the more brownie points you'll get.

Taking soundings

Most local authorities are willing to be consulted about what they'd like to see built in their area, and the subjective nature of planning decisions makes a pre-application consultation with a planning officer a really worthwhile investment, especially where design is or could be an issue. As a first step, ask your architect to draft some initial sketches of the building or conversion and its siting; it's important that you go to your meeting with something visual. These drawings can then form the basis for a discussion with your planning officer who, in turn, might consult other council officers responsible for, say, conservation, urban design or landscape. This first move can either be a face-to-face meeting or you could just send the sketches to the planning department informally, with a letter. Either way, this early tactic allows you to be able to react to the planning department's response, address any concerns they might have and be aware of what the potential problems might be. The scheme might need slight amendment or possibly a fundamental rethink. If, having reviewed the planning officer's comments carefully, you and your advisors don't want to compromise, you are not obliged to – but you could be in for an uphill struggle. Furthermore, the planning officer's comments might indicate a need to speak to others, for example, the highway authority about issues of access. Having support from authoritative architectural bodies, such as the CABE and local architectural organizations or experts, will add weight to your application, so a pre-application consultation with them can also be useful. Put simply, the ideal position is to have all parties' endorsement before you submit your application – and it needn't cost an arm and a leg.

The planning application process

Once you've filled in all the forms, got the drawings and supporting material together, and attached the council's statutory application fee, the application has to be submitted to the council. This is normally done by post but there's a move towards electronic submission of applications. The council starts by ensuring that all the requirements have been met and the right boxes ticked, and then it writes to the applicant or the applicant's agent (representative) either 1) requesting additional information or 2) just confirming receipt of the application. The acknowledgement gives the application reference number, which needs to be quoted in all contact with the council or used to look up the application on the council's website, and it also gives the name of the planning officer dealing with the application.

THE CONSULTATION PROCESS

The council then notifies all the relevant consultees (see above) that an application has been made and invites their comments and observations. By speaking to the planning officer, you can find out who is being consulted on the proposal and you can contact them directly if you want to discuss the application and iron out any issues.

Consultees' responses come back to the council within three or four weeks, during which period the council also publicizes the application by posting up a site notice (usually an A4 piece of paper stapled to a telegraph pole, gate or somesuch at the site) and/or writing to neighbours. Local newspapers also usually publish a list of applications submitted. Any response or 'representation' from the public about an application has to be made in writing (including email) and anyone can write in, not just those who are notified. Many councils have an 'open file' policy, meaning that all information is made publicly available as it's received, enabling you to see exactly what the consultees and local people have said. This opportunity is important because it gives you a chance to respond, to clarify points and/or counter what someone else has said if the application comes before a planning committee.

WHAT DOES THE PLANNING OFFICER DO FOR A LIVING?

Planning officers are dedicated, underpaid souls who aspire to make the world a better place, by helping to design the way our towns and our countryside are laid out, and who feel utterly unsupported by either central government policy or, for that matter, by us, the public. Very often they are pressured into processing applications quickly simply to meet central government targets, and in any case they won't usually look at an application in any detail for at least a month – by which time all consultees' comments should have been submitted. The officer will study the application, pass comment and probably inspect the site – they don't usually make appointments to view properties but you can request this as it gives an opportunity to discuss the proposal, figure out the officer's attitude and, most importantly, personalize your application by putting a face to the paperwork.

To further maximize your chances of success, you or your representative should monitor the progress of the application by staying in touch with the planning officer. Councils vary in their policy and practice and some have very open and accessible planning departments, but it has to be said that in many authorities it's hard to speak to planning officers but easy to speak to their voicemails, which are left on permanently. This is poor practice and all you can do is keep leaving messages. The officer should let you know how the application is progressing through the process and what kind of impression it has made – most officers will tell you if they have concerns. Again, this gives you an opportunity to address those concerns and maybe amend the application in some way.

Although an individual officer's views are, of course, coloured by their personality and tastes and by their departmental culture, their decisions are by far and away primarily informed by planning policy, policy guidance, site-specific factors that they'll have taken into account and all the comments received. This variety of factors, which can be highly contextual and particular to your application, explains why permissions are sometimes inexplicably granted or refused, even if a neighbouring application has received the opposite response. All the evidence and opinions are weighed up and the officer will then form a view on whether the proposal is acceptable. He or she then writes a report setting out all the relevant considerations and culminating in a recommendation for approval or refusal.

DECIDING PLANNING APPLICATIONS

Applications are decided in one of two ways – by committee or by delegated powers. When an application is contentious (effectively when any of the consultees or members of the public raise any kind of objection or concern), it is submitted to the local authority planning committee for a decision. Your local planning committee will be made up of a number of elected councillors – lay people, not experts. Sometimes this can work in your favour because the debating process is open to the public, but very often local committees don't have a full grasp of more complex design issues nor an appreciation of architectural diversity. Councillors are political creatures, too: in the worst cases, committees can become arenas for political manoeuvring.

If your application goes the committee route, the planning officer's report on your scheme is included on a committee agenda published a week or so before the meeting. If your proposal has got the officer's backing, all the better, but it doesn't guarantee success at committee. If the officer recommends refusal, there will be a short window of opportunity to write to the council to state your case, answer the objection and exploit the democratic and political structure by lobbying councillors.

Councils' policies on lobbying vary but, in most areas, you're allowed to write or speak to councillors on the planning committee to try to win their support. Most councils now also allow applicants and objectors to speak at committee meetings, and this is another good opportunity for you to wade in and defend your application. The time allowed to speak is short – two or three minutes – but it's often really worth doing – though it's crucial to prepare, rehearse and remember to keep your arguments clear and entertaining. You might want to use visual material (as long as it's easy to understand), and you should certainly have an advance understanding of the objections and issues that are being raised so that you can go straight to the heart of the matter and deal with those in your precious few minutes. You'll know how successful you've been because after discussing your application, the committee will vote on whether to approve or refuse permission.

When applications are more straightforward, with no objections, they can be decided on by delegated powers, which means it's the planning department (the professional civil servants) that makes the decision. In this instance, the officer's report is passed to a senior officer for checking, and he or she might well take a different view from the planning officer who dealt with the application and overrule the recommendation. And it's not just the direction of the application that can be unpredictable: the timing of delegated decisions is more uncertain, too, because, unlike committee meetings, which are regular and scheduled, there's no fixed timetable for taking delegated decisions. Planning officers often don't make up their minds about a proposal until just before the eight-week target date.

DECISIONS NOTICES

Whether refused or approved, the verdict on your application is conveyed in a decision notice sent to the person who submitted the application. Planning approvals are given subject to conditions that you must check carefully. Permission is normally subject to a three-year time limit within which to begin (not complete) work and after that time, if not acted upon, permission expires. Some conditions are standard, such as council approval of external materials and submission of a landscaping scheme. Many conditions, like these, require further approval before work begins (sometimes referred to as 'reserved matters') and this should be obtained before work starts so as not to invalidate the permission. Conditions on outline permissions can restrict the size of house or number of storeys. Occasionally, conditions require work to be done outside the application site on land not owned by the applicant and the necessary rights then have to be negotiated with the landowner concerned before the development can be built. When planning permission is refused, the decision notice sets out the reasons why the council turned it down. These reasons are written in technical planning language, the meaning of which might not be entirely clear to you. Refer to the officer's report or talk to the officer to get it translated into English.

APPEALS

If your application is refused, there is a right of appeal and the decision is then made afresh by an independent inspector (England and Wales), reporter (Scotland) or commissioner (Northern Ireland). You need to decide whether to go the appeal route or reapply to the council for a different scheme – or both; you are likely to need professional guidance in order to make an informed choice. If your original application was turned down on a point of principle, the council probably would not change its mind and so the choice is between appealing and giving up. However, if the application was refused on points of detail, reapplying with an amended scheme is likely to be the quickest route to permission – assuming you're prepared to compromise.

If you decide to appeal, you'll be encouraged to know that a large number of appeals overturn original planning decisions. But it's worth remembering that the procedure can be lengthy. More than three-quarters of appeals are conducted in writing but it is possible to end up in a hearing or even an inquiry for the most significant schemes.

Listed building consent

This is a more rigorous version of planning permission that applies to any building that appears on the government's list of buildings of historical importance. The English Heritage website explains the process of listing but essentially any building can be nominated by anybody; the nomination is reviewed by English Heritage, a statutory body, but the final 'decision' is made by the Culture Secretary in the Department for Culture, Media and Sport on EH's recommendations. For years buildings have been listed Grade I (for the very finest buildings) Grade II* (Grade two star) and Grade II (the majority of the half million or so listed buildings in England) but the decision was taken in 2005/6 to amalgamate Grades I and II*. The process is administrated in Scotland by Historic Scotland and in Wales by Cadw.

In practice, what does it mean? The process of applying to do any works to a listed building is nigh identical to the conventional planning route, except that you can't do any work at all to a listed house without consent. You can't remove internal walls or build a garage within its 'curtilege' (the developed site on which the building sits), or strip out the skirting boards. You certainly can't put some uPVC double glazing in – and you won't get permission to do so, either. You're not even meant to change the colour of your living-room walls without permission. You are allowed to breathe in the house, though.

Most local authorities are happy for you to choose your own internal decorating colours (unless the house is Grade I), but do remember that every stringency that applies to your house applies to your outbuildings, potting shed and outside dunny. Sometimes to the landscaping, too. And it's a fallacy that if the description of an outbuilding or internal feature doesn't appear on the listing description of the house, it isn't subject to consent. It is.

Decisions for Grade I houses tend to be referred to English Heritage regional offices, which will grant permission and consult on the works. For Grade II buildings, the decisions come from the local authority and are made either by the planning committee on planning department recommendations or by the planning department under delegated powers, and in fact by the department's conservation officer, a trained specialist, usually. Appropriate amenity bodies such as the Georgian Group, the Twentieth Century Society, English Heritage and the SPAB are always consulted.

Because of the intellectual complexity of many listed building decisions, answers can sometimes take longer than the statutory eight-week maximum

facing page: **Sker House, Wales.**

period. In fact, if English Heritage is involved in consultancy, the review and consultation period can sometimes be months and involve the building's owners. But the quality of the decision you'll get back will tend to be extremely variable from one local authority to another. This is because up to a third of councils don't even have a conservation officer, because they're not statutorily required to have one. They're just saving money. And when you do find a conservation officer, that's no guarantee of expertise. My own local officer runs a small conservation department, has several degrees, worked at English Heritage and has an encyclopaedic knowledge of local building styles and traditions. Hurray! But I met one recently who'd left school at

16 and got the job only because he was working as a junior in the planning department. The head of planning stuck his head round the door and asked if anyone was interested in old buildings. 'I was the only one to put my hand up,' he told me.

If he's your conservation officer, do what I've recommended elsewhere: sign up for a SPAB homeowners' course and pay for him to go, too.

If you think owning an old listed house entitles you to pots of grant aid, think again. Works and repairs to old buildings using traditional techniques, craftsmanship and materials such as lime, oak and stone can be slow and expensive. Thanks to listed building consent, such techniques, craftsmanship and materials are nearly always compulsory, not

above: **Sker House, Wales.**

an option to using concrete and plastic windows. You can ask about grants, but nearly always you will be paying for the work. It is part of the onerous responsibility you adopt when you buy a listed house – matched by the benefit of living in a building with charm and character that you enjoy. That's the deal.

Don't think that if you buy a house that isn't listed but is in a conservation area the regulations are any less stringent. Conservation area policies are set out to national guidelines but written locally, and they can be as rigorous and demanding as listed building requirements. Check out your local authority's policies on its website and while you're at it see how many, if any, conservation officers it employs.

Finally, if you're fazed and angered by the idea of someone else telling you how you can or can't arrange and decorate your own home, remember that listed building consent is for wimps. Try living in a Scheduled Ancient Monument. Francis and Karen Shaw did (see page 213). They bought a derelict peel – or defended house – in Yorkshire that was subject to Scheduled Ancient Monument Consent, which covers not only the building but its surrounding land and archaeology. Getting SAMC is nothing like planning permission. It involves English Heritage officers at every stage of the project – officers who can stop all work at any time; you need listed building consent, as well; you have to employ an archaeologist; and you might as well leave your chequebook open and file for bankruptcy on day one. Francis had no idea where the project would lead him and ended up saying goodbye to £25,000 on archaeology that yielded very modest results. We followed the build over three years for *Grand Designs* and made a special 90-minute film about it.

Trends in planning

There have been several changes to the spiritual direction of planning in the past five years, at the heart of which is sustainability – meeting the needs of the present without compromising the ability of future generations to meet theirs. Broadly speaking, planners interpret sustainability in a number of ways.

First, they like to see the environment safeguarded at a specific local level, which means tree protection and wildlife and water resource conservation. For larger schemes and for sensitive sites, councils like to see Environmental Impact Assessments – which are time-consuming and costly to commission. The ever-increasing regulations and more stringent standards applied to new building have an impact on planning applications and make securing permission a more complex process. The consequence is that you might have to carry out a great deal of investigation. This can include flood risk assessment, drainage reports, ecological surveys, contamination research, archaeological surveys, traffic studies and so on. The pressure on councils to meet application handling targets means that such investigation has to be carried out before the application is made because there would not be time once the application is submitted.

Second, they like to see the use of natural resources minimized during the 'life' of the building. This means sensitively designed drainage systems that deal with surface water on site and the use of rainwater harvesting and greywater recycling. They also want to see less car dependency, which has a bearing on proposals to site houses or carry out conversions in remote locations and on proposals that might overburden central urban settings.

Surprisingly, this is often as far as it goes. Planners aren't that interested in the more hard-core and more important aspects of sustainable development such as low-embodied-energy materials, district heating systems from biomass or geothermal, local supply of materials, the recycled and recyclable content of a building, local power generation, passive solar gain, heat recovery and super-insulated buildings. In fact, so little are they interested that these issues rarely affect the success of a planning application. Although there are quiet and subtle hints that within ten years they will.

Another welcome change in attitude is a more proactive stance towards what constitutes good design, thanks mainly to the campaigning of the CABE. Central government wants councils to publish design policies and statements; most councils can't be bothered. But a few, like Bexley for example, have produced guides and codes for what constitutes good design in their district or borough. As a result, it isn't unusual to find a tension between architects and their clients on the one hand and planners on the other over how far councils should dictate on design questions. The government's view is that design codes and policies must avoid unnecessary prescription, not attempt to impose architectural styles or tastes, nor stifle innovation, and that design should be appropriate in context, improve character and appearance and reinforce local distinctiveness.

Sustainability statements and design statements are now pretty well a standard component of any planning application, and rightly so. They explain and justify a scheme's wider environmental and contextual value but they're often undervalued.

Apart from anything else, the default position of your average planning committee is a cautious and negative one, a position not disposed to the construction of ground-breaking, sustainable and contemporary architecture, at least not when it comes to private houses. Some of the boldest house designs have to go to appeal to be won.

The implication of this is to choose your site carefully and ensure that your design respects and reflects the specific circumstances of the setting. In England, as an exception to the normal prohibition on building in the countryside, there is scope, through PPS (Planning Policy Statement) 7, for isolated new houses where the exceptional quality and innovative nature of a design provide special justification. Only a handful of such houses are ever allowed. But the possibility is tantalizing for many ambitious would-be house builders.

There are now hints that the wider landscape of Britain might be protected even more than it is. English Heritage produced a draft document in 2006 aiming to rewrite its approach to conservation of the built environment. It proposes a lot of changes but broadly speaking tries to shift focus away from the individual to the general. English Heritage wants to conserve not just isolated buildings or objects in the landscape, but adopt a wider approach that makes sense of a building or a place in its wider context. Context is a big word for them. It replaces ideas such as 'archaeological narrative' and the more strict conservation-led approaches of organizations such as the SPAB, which is a shame. Our conservation culture is probably the best in the world on account of its scrupulous and almost academic approach to how we conserve our heritage without heavy intervention or 'interpretation'. It strikes me that in widening the scope of its view and welcoming 'accessibility', English Heritage may also be softening the focus and moving towards the more continental model of 'patrimony', where buildings and places are highly interpreted for the contemporary palate and where the connectedness of ideas across landscape comes before the need to assess every building on its own merits – as is the case, for example, in Tuscany, where there is an official pressure for buildings and hilltop towns to conform to certain stereotypes. The upside might be a greater public awareness and appreciation of our heritage. The downside might be a conservation culture with less integrity and a country where new buildings are even harder to put up.

As to whether the planning system can cope with any of these changes, I doubt it can. As a culture, planning lost its way 20 years ago, moving from a polemic and design-led position to a bureaucratic one that was all about control. For planning to flourish in Britain, for it to adopt and promote 1) high-concept design, 2) sustainability principles and 3) a sophisticated approach to heritage, several things have to happen. First, we have to see planning policy decentralized and given back to the regions so that more flexible, creative and responsive policies can be written to suit the economic and social dynamics of an area. This might even help to take the heat out of the housing market by managing supply and demand more sensitively. Then we need to allow greater fusion between the separate disciplines of urban design, conservation, planning and building control and reintroduce creativity. And finally, we need to pay planners more. I've said elsewhere that £25,000 for a senior planner is a pittance for someone who has the power to change our environment so profoundly. This is an important job and we need to attract the best creative brains into the discipline, instead of letting them become hedge fund managers. But what do you think the chances are of your local planning officer earning £150,000?

Party-party
I couldn't resist just mentioning the Party Wall Act 1966. Nothing to do with planning, really, but dealing with an issue that can threaten the success or even outcome of your build just as seriously. The Act provides a framework for preventing and resolving disputes in relation to party walls, boundary walls and excavations near neighbouring buildings. So if you're building your new home and are: 1) working on an existing wall shared with another property or 2) building on the boundary with a neighbouring property or 3) excavating near a neighbouring building, you should find out whether the work falls within the scope of the Act. If it does, you have to serve a statutory notice on all those defined by the Act as adjoining owners – your neighbours. A notice must be given even when the work will not extend beyond the centre line of the party wall. As you can imagine in Britain, land of the Leylandii and Neighbours From Hell, people fall out all the time over their party walls, which is why the Party Wall Surveyor was invented. He is a surveyor whose sole job it is to help resolve disputes over boundaries. His profession is almost unique to these shores – and those I've met are very, very busy. You can find one from The Royal Institution of Chartered Surveyors (RICS) at www.rics.org.

Getting your Head Around a Building: Drawings

Some consider them a necessary evil, others just necessary. But you can't really build a house without drawings. The clue is in the word 'build': because construction is all about the layering of thousands upon thousands of tiny components that have to go together in the right order, so that when you turn on the tap, water comes out instead of gas, and when you open the door, you don't walk straight into a brick wall. Drawings are the code that will unlock a happy build. They are the language of building. People even talk of 'reading' a plan. Which means that they usually need translating for anybody outside the construction industry.

Architectural drawings communicate the ideas of an architect into a visual language that's easily understandable by builders, engineers and surveyors, if not by you. Many people building their own homes really struggle to understand architectural drawings and can only truly come to terms with what's been worked out when they see it in three dimensions. This is the Frankenstein 'OhmyGodwhathavewecreated?' moment, when the design is up and built. Which is when it's usually too late to do anything about it.

Drawings are also an invaluable tool in the ongoing design process. They're the means whereby ideas get explored in the cheaper two dimensions before being fully expressed in the more expensive and material three. Not that everybody adheres to that tenet. In series six of *Grand Designs* Peter and Christine Benjamin built an ambitious glass-walled pavilion in their walled garden in Devon (see page 272), often reverting to the less-than-orthodox method of design by construction. This is similar to the Frankenstein process and involves finding the time to build a component of the house then, if it doesn't work, knocking it down and trying again. Obviously, you can only perform this prehistorically experimental method so many times before you expire of creative exhaustion or run out of money, but, mercifully, being visually competent, they didn't have to repeat the method too often and did in fact resort to the odd sketch on the back of an envelope at times.

Call me old-fashioned, but I was schooled in the 4,000-year-old method of working it all out with a pencil first, and will, at every opportunity, prefer to resolve a design problem by drawing on a piece of paper. I find that I can consume a lot less energy and attempt countless variations to a drawing in an hour or two. And rubbers are a lot cheaper than wrecking bars. Drawing something out, no matter how badly, can help to resolve design issues and, more importantly, commits an idea to paper as opposed to leaving it swirling around your head, thus freeing your mind for more important things, like lunch.

And if you get really good at drawing, you'll fancy your chances at resolving three-dimensional issues, such as compound locking scarfing joints to repair a medieval oak beam. At which point you will truly be competent enough to sack your architect. Until then you will have to plough on trying to understand both their drawings and, even worse, their handwriting.

2-D and 3-D drawings

There are two basic categories of drawings: two-dimensional and three-dimensional. Plans, sections and elevations are the most common types of two-dimensional drawing – a plan being an aerial view or cutaway through a building, an elevation being a drawing of a wall side-on, and a section being a sideways cutaway or 'slice' through a building. You'll often see a section referenced as A–A or somesuch, and this usually refers to a particular 'slice' through the building as marked by a line from a point 'A' to another point 'A' on a plan. Plans, sections and elevations provide all the basic information about a building: a plan shows length and width; elevations and sections will show length or width plus height.

The plan is probably the most common and most basic drawing of all, and a great number of architectural problems – such as space management, aspect, use and circulation – are resolved on plan. A plan will usually take a horizontal slice through a building, as if cutting midway through the windows, looking down. It'll show the positions of walls, doors and windows, and can show details such as floor patterns, furniture and door 'swings': particularly useful for figuring out how much usable space you'll have in a room. At a more sophisticated level, a plan can show engineering and structural details, floor joist layouts, and electrical, ventilation and plumbing diagrams. And it is a good, satisfying word. Having a plan suggests that you might just have a plan.

The section slices vertically through a building in the same way that the plan makes a horizontal cut. The most important thing about the section is that it allows you to understand where the pipes go. It will also point out the room where the architect has inadvertently provided you with only 1.6m of headroom. And it will reveal to you just how many changes of level are required in a house and how deep the foundations might have to go. It may even explain how the building stands up. Just like a plan, a section has a certain raw naked truth to it, as though somebody had ripped the front of your house off and discovered you in *flagrante delicto*.

Not so the elevation, the most polite of drawings. It usually details the most refined architectural and the least structural elements of the building: the arrangement of the façade, windows and features; the materials to be used; and the height of the building (which can be helpful in understanding the scale relative to other fixed points such as adjacent buildings or trees). An elevation can also be the most persuasive of all drawings because it represents the house as it might be seen from the street. As such, a finely rendered elevation can form a powerful part of a successful planning application.

For drawings to be really persuasive, architects use perspective, axonometric and isometric projections. Big words used to describe 3-D drawings of sometimes mind-boggling complexity (at least to produce). Truth to tell, most architects now cheat and get their computer to produce these drawings by feeding it such basic information as elevation, plan and section dimensions. Or cheat by getting a hapless assistant to draw them.

Perhaps the most common form of presentation drawing is the perspective (as established by Italian Renaissance artist and architect Filippo Brunelleschi and codified by Leon Battista Alberti), which creates the compelling appearance of three-dimensional space as we see it. It's based on the understanding

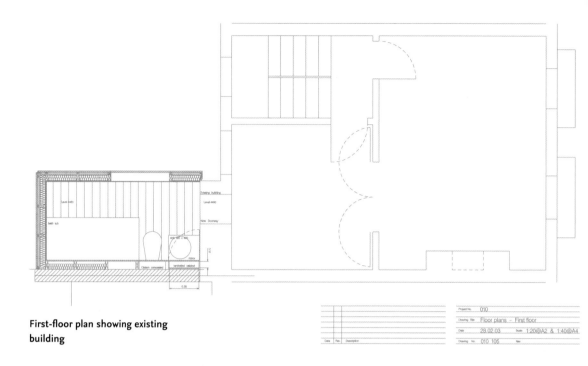

First-floor plan showing existing building

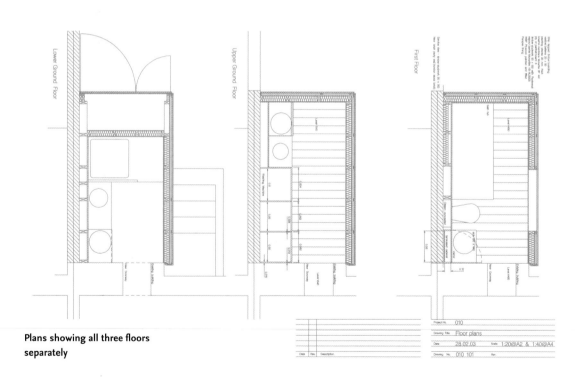

Plans showing all three floors separately

Henning Stummel's design for the spaceship extension to a London house.

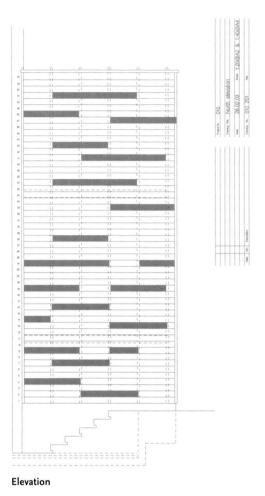

Elevation

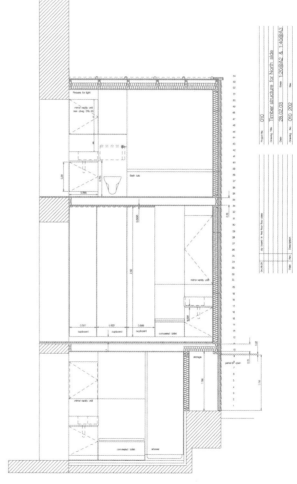

Section

that parallel lines and planes appear to converge as they recede into the distance and the principle that any line which is truly horizontal will lead towards a point on the horizon. When presenting a scheme, architects will often use perspectives from a number of different angles and viewpoints in order to present or sell the ideas in the best way, including low viewpoints where none of the lines on the paper seems parallel as they taper dramatically in all directions. This form of extreme perspective usually represents an architect at his or her most desperate to inject some excitement into a scheme. And as a result, planners will often ask for computer-verified views of a project, especially of larger developments, on the premise that the computer cannot lie.

The more exotically titled axonometric and isometric drawings are, sadly, a little more dull than the conventional perspective, being more mathematically formulaic.The axonometric drawing extends walls up from the plan at 45°. Isometrics are similar to axonometrics but generally are more realistic; the horizontal edges of the drawings are drawn at an angle of 30°. Oh, what a letdown. The only advantage to these drawings is that they can be measured in every plane and are thus a bit more useful to the builder. In practice, they're much more useful to the interior designer who needs to calculate wallpaper lengths and figure out where to put the pictures. And they can be drawn by the talentless. Or by interior designers.

What, when?

It's important to understand the specific role each type of drawing best serves so as not to insult your architect, avoid potential wallpaper disasters and win planning approval every time. The militant core – plans, sections and elevations – are usually referred to as planning drawings if they're furnished with the minimum of dimensions and look pretty; and working drawings if they're covered in figures, scribbles and cement dust. And there's a difference between a drawing showing compliance with the building regulations and a drawing that can be used for construction purposes. The former only show compliance with the building regulations; they don't show dimensions and construction details.

The hardest drawings of all to do, perspective drawings, are pretty well useless on site, though occasionally they can be pressed into service to explain a complicated arrangement of elements – when many layers of different materials come together at a corner, for example. But perspectives do get across how a building looks in context with other buildings, so they're useful at the planning application stage when the concept is presented and 'sold' to planners or their planning committees.

Scale

It would be a waste of time and paper and space to draw whole buildings at the same size as they appear in reality, wouldn't it? So architects use different 'scales' for their drawings. Drawing a building, or a detail of a building, to scale is a measured way of reducing all the information about a building so that it fits on a conventional-size piece of paper. Generally, the more complicated the information, the larger the scale used. A detailed drawing of a roofing edge or the design of a window cill may be drawn at a scale of 1:10; this means that the real thing is actually ten times bigger than shown and, conversely, the drawing is ten times smaller than the real object. I hope you're still with me.

Common scales for plans and elevations are 1:100 (that is, one-hundredth of the real size), 1:50 1:20 and 1:10. For block plans or site plans (part of your planning application), your local authority may specify larger scales, such as 1:200, 1:500 or 1:1250, the general rule of thumb being that a scale is always chosen relative to both the level of detail and the size of area covered by the drawing. The miracle of scale drawings is that they not only allow the designer to draw his or her building accurately, they also allow you to take measurements from the drawing using a scale rule. You can buy a scale rule in any good design supply shop or office stationers. Nowadays they're calibrated in sensible metric scales like those I've mentioned. Go to America and you can only buy imperial scale rules for measuring in inches and feet, which were last available in shops in the UK in 1978, arcanely measuring 1:24 or 1:6. Useless unless you're importing an American kit-house – which is exactly what some of our Grand Designers do: in Oxfordshire Chris Ostwald built through the summer of 2006 a reproduction eighteenth-century mill constructed on a Douglas Fir timber frame imported from New England. Not only did he have to revert to an eighteenth-century scale ruler, he also had to remind himself how to measure 17/24ths of an inch and calculate the square yardage of the floor plan.

Not that we are entirely rid of the medieval system of measuring based on bits of the human body (an inch is the length of the upper thumb, a yard the distance from hand to neck – useful for measuring cloth and string – and a foot is a foot.

Although horses are inexplicably measured in hands). When reading drawings, you may find references to sheet timber such as plywood dimensioned at '1220 x 2440mm'. Impressive, eh? Millimetre measurements – or dimensions, as architects and builders will call them – are designed to impress because they suggest accuracy. In truth, a plan of a building covered in millimetre references is impossible to gauge properly because there are always too many noughts after the numbers. And as for the plywood, well, 1220 x 2440 is actually 4 feet by 8 feet, a good old imperial size that lives on cloaked in zeros.

But whichever measurement system you use and whichever rule you buy, you'll find it difficult to use on account of the fact that every drawing you ever pick up will bear the terrifying words DO NOT SCALE. This is your architect's insurance disclaimer and the ultimate defence behind which he or she can retreat. The millimetres may befuddle you, the interchangeable words isometric and axonometric may confuse and distract you, but the phrase DO NOT SCALE will stop you in your tracks. It is worth ignoring it, of course, just to see how inaccurate and millimetre-imperfect the photocopied, stretched and distorted drawings that you now have truly are. And builders often have to resort to scaling from drawings because measurements – sorry, dimensions – are often conspicuously absent from them.

Virtual modelling or balsa wood?

The computer is such a design tool now that on graduating after seven years of study, young architects can expect to spend the next ten years

sitting in front of a screen drawing gutter details for an architect in his fifties who, by the way, will grandly announce to his clients that he only uses a pencil. Such is the life of the CAD monkey – CAD standing for Computer Aided Design – because the partners in the practice are Crap At Detailing.

CAD is playing a more and more significant role in the production of all types of architectural drawing, from full-colour presentations to highly technical plans. Many complex buildings – such as the Gherkin in London and almost all of Santiago Calatrava's work – have been made buildable only thanks to complex CAD engineering software. Computer simulations can be used to model highly complex engineering structures through time, so that, for example, a bridge can be 'shaken' in high winds on the computer screen to test its strength. CAD drawings are essentially built in layers, allowing specific layers of information, such as wiring diagrams, to be printed separately or collectively, so that several service elements can be overlaid to check that they'll all fit in the same conduit, for example. CAD drawings are also relatively quick to produce and easy to alter (provided the operative is skilled) and therefore very cost-effective.

And by plugging bits of software together, an architect's computer can provide the planning drawings, technical details, engineering drawings, a bill of quantities, a groovy fly-through animated sequence and, of course, their invoice.

But there's still a place for hand drawings. Most architects develop their own drawing style: one I know uses a deliberately blunt pencil when sketching out proposals to a client in order that the edges of the concept remain 'a little blurred' – a useful device when you don't necessarily want the client to be focusing too early on the detail rather than the general massing and arrangement of the building. Sir Norman Foster does most of his work with a pencil, but then with 200-odd CAD monkeys labouring for him, he can afford to. Will Alsop and Zaha Hadid both develop their work as paintings as part of the early design process.

Despite all the cartwheels and plate-spinning that computers are capable of, it is, ironically, the pencil drawing that often wins over the planners' hearts. A beautifully rendered drawing is, of course, just a few marks on a piece of paper, but it has value and character – which seem to echo the value and character of a built structure in the real world. Both are the work of human beings rather than machines.

If the pencil is the architect's essential 2-D tool, the cardboard model has to be its 3-D equivalent. It's impossible to overvalue even the most basic sketch model for helping you, the builder, the planners and even your architect to understand the building you're going to put up. Insist on more than one model, no matter how ropey they are, each one built to a different scale, so that you can appreciate the building in its context as well as in its detail. It is quite astonishing how, despite reams of drawings and nights of wrangling, there always remains one detail of the building that no one has really got their head around. A model will help you to sort it out. More so, perhaps, than a CAD model or fly-through, which, though highly seductive, can conveniently pass through walls and climb staircases supported on thin air. In many ways, the CAD model can be the ultimate deceit of what you're getting, the cardboard model the ultimate 'trial go' of the real thing.

previous page: **Meredith Bowles of Mole Architects' model for his own house. The design was born of his fascination for agro-industrial buildings, sheds and silos in the surrounding Cambridgeshire countryside.**

Building Regulations

I really had hoped to keep this section short, because building regs are not the sexiest subject.

Having said that, I owe a debt of gratitude to every building control officer (the people who apply the regs and check the work on site) I've ever worked with. Unlike planning officers who are unlikely to return your call until sometime in 2068, BCOs ring back the same morning and then turn up to look at your foundations next day. How's that for service? Forget hospital waiting figures or A level pass results. If the government wants to big up its reputation, it should parade around Britain's building sites trumpeting how blindingly brilliant building control departments are. So, to BCOs around the country, this one's for you.

Don't, however, feel that you need to read beyond this point.

So, what are building regulations?

Building regulations are there to ensure that your home is constructed properly and doesn't threaten your health and safety. They have a fundamental effect on your home design, whether it's a new home, an extension or an alteration. The principle areas they cover include fire safety, ventilation, hygiene, drainage and the conservation of fuel and power. This is all summarized in a detailed set of documents (called approved documents), which are generally well written and well illustrated. I wouldn't suggest that you invest what is a fair amount of money in these documents, not least because they're constantly being updated. But you can view them on the website of the Department for Communities and Local Government (DCLG), which is responsible for building regulations. Go to www.communities.gov.uk.

The approved documents include:

Part A Structure

Part B Fire safety

Part C Site preparation and resistance to moisture

Part D Toxic substances

Part E Resistance to the passage of sound

Part F Ventilation

Part G Hygiene

Part H Drainage and waste disposal

Part J Combustion appliances and fuel storage systems

Part K Protection from falling, collision and impact

Part L Conservation of fuel and power

Part M Access to and use of buildings

Part N Glazing, safety in relation to impact, opening and cleaning

Part P Electrical safety

Regulation 7

Materials and workmanship

The regulations are administered by local authority building inspectors, the heroes of this chapter, and they usually come under the control of the local planning authority (though in London the arrangements are slightly different). But there are also now alternatives. Instead of a local authority building inspector, you might find your site visited by an approved inspector licensed through the DCLG, or an approved inspector from the NHBC (National Home Builders Council).

In fact, as someone building your own home, you could make the choice between a local authority inspector, an approved inspector from a private company and using the NHBC. Follow the advice of your architect; in certain areas you may find that one route is more efficient than the others. It's certainly worth getting fee quotes from all to begin with, and remember there may be other benefits of using an organization like the NHBC – which offers a building guarantee in addition to simply checking the work to see that it complies with the building regulations.

When do the building regulations apply?

The building regs will apply to practically all the work you'll be undertaking. If you look at the headings of the approved documents on page 259 you'll be able to figure out that they cover pretty well every stage of construction. Even if you're converting an existing building into a new home, the building regulations will apply, though there are certain exemptions when working on a listed building, such as reduced need to comply with Parts C, E, F, L and M (double glazing, for example, may be forbidden, and walls and floors may need to be left without damp courses to allow moisture to transpire through the building's structure as intended when built). It's often difficult to establish when works to a historic building are simply repairs and when they are alterations. Speak to your local building control department or your conservation officer before starting work to figure out if you need building control approval.

If you're making changes to an existing place that isn't listed, then compare what you're doing against this checklist:

· Repair work: Building regulations normally do not apply to repair work where elements are replaced with the same, or similar, materials by way of repair.

· Examples of exemptions are: where the replacement work is extensive, for example, of a flank wall or a bay or the whole of a roof where a defective timber beam is replaced by a steel beam.

· Underpinning: Building regulations apply to underpinning (unless the building or extension being underpinned is exempt).

· Building regulations apply to replacing roof tiles or slates unless with the same or similar materials.

· Loft conversions: Where a room is formed in the roof space, whether it is a habitable room or not, whether it has a stairway or not, the building regulations apply.

· Replacement windows: Building regulations apply to replacement windows and to some replacement doors.

· Installing new fittings: The building regulations apply to the installation of some fittings, including basins, WCs, baths, showers and sinks, unless they replace existing fittings in the same room. They will apply to forming a new bathroom or a kitchen in an existing room.

· They apply to the installation of solid fuel, oil or gas-burning heating appliances.

· Structural alterations: The building regulations apply to structural alterations, such as where an opening is formed in an existing load-bearing wall (some timber-frame walls are load bearing) or an existing opening is widened.

How do they check up on you?

You need building regulations approval like you need planning permission. The difference is that planners tend to leave you alone once they've granted permission, until somebody complains about the volume of your builder's radio or the height of the radio-telescope you've constructed in the back garden. Whereas building inspectors hang around your building site all through construction, checking the depth of your concrete and blagging biscuits from your site hut.

Building work is usually inspected as it's carried out: an inspector from Building Control will, for example, want to see your footing trenches when dug, to check their depth, and then again when back-filled with concrete. They'll want to see your drains, too. And your damp-proof course. Who said life in local government was mundane?

Out of the ground, they will then want to examine the building at various structural stages, to look at floors and roofs, principal beams, steel components of the building and reinforced concrete. You can see that there is a strong engineering bias to their involvement.

Because it is so key that your inspector sees at first hand certain precise moments in the build, it is vital that you or your architect or builder let them know, a few days in advance, when to come. Once they've made a visit at a key stage and providing they're happy, they'll issue a completion certificate. If they're not satisfied with the work, they'll tell you why and help you to resolve the problem.

It's worth remembering that just as an architect's signed-off certificate for a separate phase of construction is sometimes a mandatory requirement for the release of a block of funds from a mortgage company, so Building Control completion certificates can be helpful in securing specific insurance claims and in meeting insurance company conditions. And if you ever come to sell your house, you'll almost certainly be asked to produce a completion certificate for work done. If you don't have one, you may have difficulty in selling your home, but there is at least a procedure of regularization to help to resolve this problem.

It's worth mentioning that building inspectors do not take any ultimate responsibility for their actions in checking certain key stages of work, which is why you need building insurance and your architect to carry professional indemnity insurance (see page 271).

Stuff the regs cover

NEW HOMES

In the case of a new home, the construction must be structurally stable, keep the water out and retain heat to a certain standard. Regulations are also being introduced to cover airtightness: the prevention of heat loss through gaps in the structure.

WINDOW AND DOOR OPENINGS

There are regulations controlling the size of windows and the amount of ventilation for any habitable room. The standard of double glazing is also now controlled and there are requirements to provide air vents as part of the structure (even though they should not always be there: heat-recovery ventilation systems need airtight rooms to work efficiently and are not provided for as yet in the building regulations).

The regs also now apply to replacement glazing. That means the complete replacement of one or more windows, rooflights, roof windows or doors that are at least half glazed. You'll need to involve either FENSA or Building Control. FENSA (The Fenestration Self Assessment Scheme) has been set up by the Glass and Glazing Federation. One of the approved documents (L) gives guidance of the new insulation values for glazing, which require a higher standard of insulation than can be achieved with basic double glazing. You'll also find that where glass is low to the floor or ground, there's a requirement for it to be toughened or laminated.

FLOORS

Floors need to be able to resist moisture rising from below. They also need to be insulated and, in certain cases, they'll need to be sound-insulated, too.

DRAINAGE

The regulations cover both foul and surface water drainage, above and below ground. They control the types of materials that can be used, access for maintenance (the number and size of manholes, for example) and the fall or slope of the whole drainage system. The whole works, in fact.

HEATING SYSTEMS

You no longer just change your boiler without building regulation approval. The regs not only cover the type of boiler you install but also its efficiency and the type of fuel it runs on (whether oil, gas, biomass or fossil solid fuel). They also cover the combustion and flue arrangements. A lot of the headache of compliance can be overcome by making sure that you employ a properly qualified plumber, registered with the appropriate authority for the boiler's fuel. For example, if it's a gas appliance, the person working on it should be Corgi registered (Council of Registered Gas Installers); if oil, they should be registered by Oftec (Oil Firing Technical Association).

FIRE AND MEANS OF ESCAPE

The regulations specify the maximum distance a person can travel before getting to a safe area or out into open air. This can affect the sizes of windows (to provide a means of escape, they must be at least 0.33m^2 and at least 450mm high and 450mm wide) and even building layout. The building regulations will also normally require smoke alarms (often mains powered) to be fitted to new buildings and to existing buildings that are extended, have their lofts converted to living accommodation or undergo material change of use.

DISABLED ACCESS

The regulations attempt to ensure that any new home design, or a conversion of an existing building into a new home, makes life easier for the disabled or partially disabled individual. This includes all reasonable access to the house from where a car would be parked, either on or off the plot, to the front door; and access within the house. On the ground floor the widths of doorways and corridors are controlled. Obstructions such as radiators need to be carefully located. Heights of switches and sockets are controlled, as are stairways in split-level houses, and a WC should be provided on the ground floor. There are concessions for town houses.

LOFT CONVERSIONS

Building regulations apply to loft conversions and are particularly tight. This is because most loft conversions increase the number of storeys of the house, stressing the structure and creating more complicated means of escape in case of fire. Alternative external escape routes are normally made through a window rooflight or external balcony and building regulations will require protection such as self-closing doors, fire doors and smoke detectors. The ceiling under the loft will also need to give adequate fire resistance to protect the new floor. And the attic joists will need to be upgraded for their new role as attic floor, taking the extra load of an additional storey of furniture, bathrooms and so on. This may require steelwork. And, indeed, the whole house may need to be strengthened, re-engineered and underpinned. Finally, the roof space will need ventilating and insulating. And all these works come under the strict supervision of building control.

Steps to take to make your home safe and sellable
You're going to need building regulation approval. So employ an architect to help you get approval. The documents are far too mind-boggling to research yourself.

Get planning permission first. There's no point producing drawings and documents for Building Control, only to have your planning application turned down.

The chances are that you'll need structural work approving. It's a requirement of your local authority Building Control Department that calculations are submitted by a fully qualified structural engineer. Most home designers, architects and surveyors won't be qualified, so buy the help of a structural engineer. Your architect will be able to recommend

someone but you should employ them directly.

The application should be made on your behalf by your architect, surveyor or engineer as your agent (you remain the applicant). The application usually consists of a package of drawings and notes showing basic specifications to comply with the regulations. Note that building regulations don't cover electrical works, decorations, kitchens, bathrooms and so on. So don't expect a set of building regs drawings to be any kind of indication of total construction cost.

The application will be checked by the building inspector, who is usually the same person who'll visit the site during construction. It usually takes about three to four weeks for the inspector to look at the application and then come back with any queries – depending on where you live. With most applications there'll be a list of queries fired back at you that will need answering by your professionals. Generally, after this stage the application is approved somewhere around six weeks from the date of the original submission.

You're entitled to start the work once you've submitted the application but I always advise people to wait until a full approval has been given before starting. The building inspector will send your agent an approval notice together with cards that need to be filled in at various stages of construction. Get a copy of the approval for your files: it's an important document and you'll need it if you come to sell your home.

Remember, the building inspector, despite checking the drawings and visiting the site, does not actually take any responsibility for the work carried out, so you still need your own architect or surveyor on the job.

Once works are about to begin on site, it's the builder's responsibility to give the building inspector suitable notice (usually a couple of days) before starting work and before certain key stages of the build are ready for inspection. The building inspector will then visit the site at those key construction stages (such as excavation of foundation before casting of concrete, damp-proof course, floor slab formed, roof structure and so on). The inspector will also usually carry out a test on your drains.

You must make sure that a final inspection is carried out by the building control officer and that this inspection is confirmed to be satisfactory in writing. Solicitors now want to see this document whenever a property is sold and it's particularly important in the case of a new home or an extension to see that the work has been approved, checked and signed off by the Building Control Department.

Buy your building control officer a pint for being so helpful.

What's the latest news from the exciting world of building regulations?

Well, truth to tell, it is quite exciting. Biggest on the agenda is the government's drive to save energy and CO_2 emissions, which is going to result every few years in the ramping up of Part L, the bit governing the thermal performance of buildings. Personally, I feel that we ought to be super-insulating our houses right now to the standards that are going to be imposed in ten years' time, rather than be softly led in gentle stages. The reason change is so slow is, of course, lobbying from the construction industry, which doesn't want more costs added to its buildings. But as Part L also

rightly points out, it isn't enough to insulate our homes, we must also look at how leaky they are. The result is a new set of directives insisting on properly ventilated but airtight buildings: 'build tight and ventilate right', though as I've said, not enough consideration has yet been given to how heat-recovery systems (machines that transfer up to 90 per cent of the heat from stale air in a building to fresh incoming air) can be accommodated in the regulations. Use of resources is also an issue that has now been adopted into the Sustainable Building Codes: how to deal with and store surface run-off water, for example.

Another significant change is coming about as a result of the Egan Report on Modern Methods of Construction. In order to deal with the shortage of on-site construction skills that we currently face, and to achieve a higher quality of construction, it's inevitable that we will see more and more off-site fabrication of houses and, as a result, more prefabricated house systems using timber-frame and masonry components. And the building regulations, as well as planning guidance, are reflecting this trend. In the housing market there's a move away from high-embodied-energy technologies such as concrete towards timber- and steel-framed designs and an emphasis on minimum waste production on site. All in all, we're moving away from Roman methods of construction and beginning to build our houses more like the efficient machines they should be.

facing page: **Andrew and Lowri Davies' Carmarthen House.**

Managing your Project

There are myriad ways to manage your build. From my own experience I think the least stressful route is to go bumming around the Far East for a year while it all happens. I only say this because every building project I've ever taken on has inexorably sucked me in like a vortex. No matter that I've employed a builder, project manager and architect, my time-management skills are so non-existent that I end up spending the entire working day hanging around my own build like some kind of adolescent groupie. Work goes out of the window, copy deadlines pass me by and I find myself making all kinds of excuses for why I'm not answering the phone. More of a building junkie than a groupie...

Some people, and how I envy them, aren't in the vaguest way excited by the idea of building anything themselves, or supervising anybody else building, for that matter. They can happily assume the role of cool client, stand back and enjoy the process from a distance, evaluate every decision and take time to savour the way the design grows on site. Two of the most exemplary clients I ever met were Denise and Bruno Del Tufo, who converted a Lutyens concrete water tower with the help of architect Derek Briscoll, an extraordinary project that was broadcast in spring 2006 (see pages 8–9). Denise and Bruno had trained as artists, which perhaps gave them a subtle advantage over the rest of us. Going to art school, contrary to popular opinion, doesn't involve joining the Communist Party and consuming all the solvents in the printmaking department. What you learn there is a small panoply of intellectual as well as creative disciplines: an ability to reflect and to coolly analyse, to make decisions and then act on them. It struck me during filming that here was a couple who resolutely refused to pick up a hammer or even a bill of works and who, instead, were concentrating on making the thousands of decisions that clients need to make, all with an extraordinary air of calm and enjoyment.

Not that there is only one way to build a house. Just that if you ask me which way I'd go, it would be firmly in the footsteps of Denise and Bruno. But some choose to project manage themselves and a word of warning here: project managing a build is nothing like project managing a sweet shop. Or the IT department of Rentokil. It's more like project managing the construction of an intergalactic rocket. You might assume that having put up a conservatory, you're perfectly skilled to run a build. At which point I have to interject and say that you're bonkers. A building, like a rocket, is a machine constructed from hundreds of thousands of odd separate components from hundreds of suppliers in dozens of countries. These odd bits have to be designed, specified and ordered and brought together at the right moment and at the right place, then they've got to fit together so that the whole damned crate will fly, which is called sequencing. It's probable that the people doing the assembly might not be as skilled as you'd like, what with a shortage of labour in the construction industry. And you'll even be asking skilled tradesmen to perform unfamiliar operations because you're probably not going to be building a bog-ordinary noddy house but a Grand Design, with underfloor heating and heat recovery and category 5 cabling. So you're going to have to be a fabulous man-manager. Hey, and guess what? The likelihood is that only half of your team have ever worked together before. They've somehow got to work together and around each other. And to cap it all, you're not building your project in a high-tech factory, with parts bins, clean benches, heating, good lighting and a canteen. You're building it in a field.

All in all, building a house is much harder than building a rocket. It's got to survive longer than one vertical flight; it's got to last through thick and thin and snow and gales – and last for a hundred or so years, at that. I think it's akin to building a rocket in fourteenth-century rural Albania. I seriously believe this, which is why I am always staggered in *Grand Designs* when anybody does it successfully.

Peter and Christine Benjamin did it successfully when they built their modernist pavilion house in their Devon walled garden, a project we also screened in spring 2006 (see page 272). Perhaps that's because they're both retired and so were able to devote 300 per cent of their time to it. They were also wise enough to put together a building team comprised entirely of local tradesmen who mainly knew each other. But ultimately I felt that the house was not as perfect as it might have been had they kept an architect on full-service contract to help to resolve some of the design changes.

Tom Perry, who appeared in series two of *Grand Designs*, project managed his own home over more than three years of construction. He confided to me that it probably took the first year for him to get a grip on the job, so he really did sponsor himself through an apprenticeship of hard knocks to get what he wanted and, in my view, almost certainly

took longer to finish than he would have done employing a good project manager – which, of course, is what he is now.

The truth is that the less professional help you get, the more likely the project is going to be compromised, one way or the other. Without good management, it's going to be delayed; or over budget; or the quality's going to suffer; or the design will be compromised; or detail decisions won't get the kind of attention from you that they deserve; or you'll find yourself highly stressed and deeply unhappy doing a job you don't quite understand. As a client, one of the most important roles you have is deciding how the project is going to run. Get it right and the whole experience can be a real pleasure. Get it wrong and it can wreck lives, marriages, financial security and every dream you've had for the place. Not that I want to put you off.

How many options do I have left?

1. You can always insist on project managing yourself, citing your experience/determination/charm/foolhardiness as reasons why you should. It means typically that you would deal direct with the architect, builder and all suppliers and subcontractors. Good luck.

ADVANTAGES

This route will save you professional fees; lines of communication will be simple; and, of course, it provides a high degree of 'client involvement'.

DISADVANTAGES

Fundamental design mistakes can be made during the construction stage that compromise the design. This is usually as a result of lack of experience. A considerable amount of time is also required, usually involving a sabbatical from the day job. It also puts all sorts of pressures on partners or families because there's little time to do anything else other than to manage and organize the project. Think of Tom Perry.

2. The traditional arrangement is that between the three parties of architect, builder and client. At one time the architect would take the lead role not only in terms of design but also procuring builders and managing the project. This role has been eroded over the years and nowadays it's more usual to find a quantity surveyor or a project manager running the build.

ADVANTAGES

Provided the architect has decent management skills, he or she may be in the best position to lead the professional team and ensure that design consistency is maintained throughout – they understand the project through and through. This triumvirate arrangement is also a well-established and understood 'route' within the industry and can achieve excellent finished results. But do remember that all instructions to the builder should be made via the architect.

DISADVANTAGES

You'll have to pay your architect more for this service and he or she may well turn out not to be as gifted a project manager as they are a designer. It's also rash to rely on your architect to additionally cost the project. Get the builder to provide firm quotations based on detailed drawings, or better still employ a quantity surveyor.

3. If you're unsure about your architect's abilities on site (and there's no reason why he should be a brilliant project manager – that's not why you employed him, presumably), you could employ a building firm that then provides a site and/or project manager for you.

ADVANTAGES

This arrangement is simple and relies on the expertise of the company. It's what Bruno and Denise did (page 266). And it relieves the pressure on both you and the architect to focus on design issues. You don't need to worry about specialist subcontractors, who are all appointed by the main builder.

DISADVANTAGES

This is not a good option if the building firm is second-rate. Nor if they decide to retire or switch the manager halfway through the project. Worst of all is that the manager is not answerable to you but to his bosses and may attempt to conceal mistakes and problems from you. You'll also find that firms

are reluctant to devote staff full time to a project –
so you're almost better off making sure that you've
got an experienced good site-manager who's really
a chief builder, and persuading your architect to
take a healthy interest in the project. And you may,
of course, be paying not only for the manager but
also the firm's mark-up on his salary and on all of
the subcontractors' bills as well.

4. The best team arrangement is, in my view to
employ an architect, an independent project
manager, and a builder or set of subcontractors.
This is known as contract management.

ADVANTAGES

You get the advantages of an independent manager
working directly for you, not the builder, and you
only have one person to deal with on the ground.
They'll deal with the architect on constructional
day-to-day issues and they should pass on all their
subcontract bills direct to you for payment at trade
prices. And you get the advantage of a clutch of
subcontractors whom he knows, who work well
together and can 'dovetail' around each other
during the difficult second half of a build when
things get complicated, none of whom you have
to employ or worry about.

DISADVANTAGES

You're paying a lot of fees this way, though arguably
your manager's fee is recouped by paying reduced
subbie prices. And this is a relatively 'hands off'
position for you to occupy as client. Your manager
may also be running more than one site and not
available all the time. Make sure that they are. And
ensure some active involvement for your architect
to maintain the design integrity of the project.

5. Or you could go the 'design and build' route,
used by many home builders. You go to a design
and build company that will take you through the
whole process from start to finish, provided that
you have a site.

ADVANTAGES

This is a one-stop shop solution, simple, quick and
usually cost-efficient. The company will also obtain
planning and building regulations approvals for you.

DISADVANTAGES

The firm will almost certainly have a set of standard
home designs for you to choose from, so the design
quality and bespoke nature of the project will be far
less personal. You usually end up getting a house,
not architecture. And it'll be a 'package', in a build
technology that suits them but perhaps not you.

Should I bother with a contract?

All human relations are built on trust. Without it the world would fall apart. For that matter, every lawyer I've ever spoken to about contracts says that if you have to refer to them, it means that the relationship has already broken down, at which point a contract is then useless because the only person who'll ever profit from it is the lawyer. It's a crazy world.

But contracts are mighty fine things to have on building projects, if only because the process is so nightmarishly complicated and involves so many different parties and trades, each of whom needs the precise scope of their works spelled out for them in advance. An appropriate building contract sets out a number of important issues including payment, insurance and, importantly, what to do if everything goes wrong. You need to rely on the advice of your professional advisor in suggesting the most appropriate type of contract for your particular project. This advice will depend on the size, value and complexity of your home build.

The most commonly used contracts are those produced by the JCT (Joint Contracts Tribunal), which is a collection of various organizations involved in building construction. The JCT produces three types of contract, starting with the very simple JCT Home Owners Contract, which is really designed to cover alterations and extensions where the services of an architect or contract administrator aren't retained.

There's also the JCT Minor Works Contract, which is good for simple contracts where there aren't too many specialist suppliers and too much complicated architectural detail. JCT also produces an Intermediate Form of Contract more suited to the complex, higher-value home project. Both Minor Works and Intermediate forms require a contract administrator (usually an architect) to be employed. Some organizations lending money require that an architect or surveyor acts as a contract administrator and has a full involvement during the build. This may dictate the build route that you select.

Any properly set up contract will have a 'retention' clause – indicating a sum of money held by you, the client, to cover defects that may occur within six to twelve months following completion. This is often referred to as the 'defects liability period' and for a new home the sum in question is typically around 2–5% of the total construction cost). So monitor the building carefully during this period, making a list of all minor issues so that they can be dealt with by the builder in one go at the end.

Do remember that a contract needs to be supported by a good set of contract documents; this is discussed in more detail in Managing Money (see page 274). A building contract has a series of provisions to cater for something going wrong and it is in everyone's interest if supplementary documents, plans, sketches, on-site notes and even a build diary are kept as records throughout the process. These documents prove invaluable if you have to go to a tribunal or court to resolve a dispute. For that matter, at the end of a project you should make sure that you get hold of a set of drawings recording the project 'as built' and ensure that you have all appropriate manuals from subcontractors and suppliers. Get, also, all the approval certificates from your architect (to show satisfactory completion of works) and your local authority (any reserved matters approvals from the Planning Department and any Building Control approvals). There might also be approvals, agreements and certificates to obtain from the Highways Department of your council or the local fire officer.

Getting insurance

Insurance is about calculating risk and one thing that has always appealed to me about building is that it is a very risky business. Setting aside the fact that more industrial accidents happen on building sites than anywhere else, constructing yourself a new house has to be one of the most financially – and emotionally – risky things you can do. Just think: you take all your money, you liquefy every asset you have, you take out an absolutely HUGE mortgage that you can ill afford to pay off, all on the never-never, and you put it into the kind of project that you probably have no personal experience of. A project that is more than likely an experiment not just for you but also for your architect because, after all, every new building worth its salt is something new that has never been seen on the planet before. As I say elsewhere in this book, a proper piece of architecture is its own prototype. If that doesn't put you off managing your own project, I don't know what will.

As I'm in a doom-mongering mood, I thought it would be enjoyable to list all the things that can go wrong in a build. This is something I normally like to do just before the commercial break in an episode of *Grand Designs*. In the average programme they give me only 30 seconds before telling me to stop whingeing; if allowed, I'd consume the whole three minutes of ads banging on about overruns and overspends and too much overexcitement. The world of construction, populated entirely as it is by optimists, needs the odd voice of pessimistic reason.

Here, then, before you draw up a risk assessment, is a sanguine assessment of all the risks you could ever hope to encounter.

The risk of your professional advisors making mistakes and the consequential losses caused by this. For example, if your structural engineer designs the wrong type of foundation and it later cracks up and your house falls over, or if your architects get the specification for the cladding wrong and it peels off a year later. Professionals are only human and that's why they all cover their backsides with professional indemnity insurance – to insure them against making mistakes. So it's important that you make sure that your professional advisors carry the right kind of professional indemnity insurance and that it's valid and of sufficient value, bearing in mind the value of the project. To be frank, it's often difficult to bring this subject up, particularly at the beginning of a project when everybody is brimming with enthusiasm. But you have to put a grim expression on and do it.

The risk of you or your contractor being sued for damages. Contractor's All Risks insurance will cover the builder against any injury he might cause to individuals or any damage to property, including third parties. The same policy usually covers for loss or damage to equipment on a particular project. This may include any unfixed materials that are stored on the site belonging to the contractor. It's important to make sure that this insurance is valid and you need to make sure that the contractor indemnifies you, the employer; it's usually a contract requirement that the policy is in joint names. This protects you in that if an injured party brings an action against you rather than against

above: **Peter and Christine Benjamin's
modernist pavilion now sits in a reconstructed
period walled garden.**

the contractor, the latter has agreed to carry the consequences of the claim. In the same way, if the third party sues the employer, then you (the employer) can join the contractor as co-defendant or bring separate proceedings. The other very important requirement is that you make sure that there are adequate resources to meet any such claims, that is, the cover is appropriate. The contract, therefore, usually requires insurance cover to back up indemnities.

The risk to the building or structure during construction. Most contracts contain provisions for the insurance of the actual works. In the case of alterations and extensions, it's usually you, the employer, who needs to take out this insurance; in the case of new works, it's the builder or contractor who usually takes out the appropriate insurance.

The risk of your builder damaging neighbouring properties. This type of insurance is probably more appropriate to a tight urban situation where the risk of damage is greater, but it is also very useful if your site is next to any kind of infrastructure, such as a main road or an electricity substation.

The risk of your builder going bust or just disappearing with your money. This insurance may take the form of Performance Bonds. These cover the scenario where a builder goes bankrupt during the course of the works. Insurers will look very carefully at the risks and the track record of the contractor before offering terms. In my experience, these forms of insurance are rarely used in one-off home construction because they're expensive and to some extent they indicate a lack of trust between employer and contractor before the whole project starts. In any event, most building contracts only

demand payment for work that has been done; they don't specify payments in advance. So you, the employer, are usually fairly well protected. Having said this, if a builder disappears halfway through the work, it's usually a long-winded and difficult path to get the project back on track.

The risk of your marriage splitting up, getting a stress-related hernia, losing all your money and losing all common sense. These risks are rather real. In fact, not so much risks as likelihoods. As a result, no company is prepared to insure against any of them. So, tough.

Managing Money

Stabs in the dark

One of the biggest guessing games in construction is working out how much things will cost. Not when you've got a full set of detailed drawings, of course, because you can get a quantity surveyor to cost the project down to the last roofing nail. But at the early stages: when you kind of know what you want but have, as yet, little to show for it.

Most people will ask their architect to bear in mind a fee while he or she is producing sketch and planning drawings. They're then horrified when a builder costs it at twice the price, which is not unusual. In fact, in every series we film, half of the projects have required some kind of redesign in order to fit the budget. The trick is to try to make sure that by the time you get to the planning process you have drawings of sufficient detail to be able to get a sensible costing, not from your architect but from a decent quantity surveyor.

A good quantity surveyor is invaluable, partly because they can find ways to save you money (by suggesting alternative build methods or materials) and partly because they are independent of the process. They're not as excited about the project as either you or your architect. Don't expect them to be. The whole of the construction industry operates under a miasma of optimism. Every architect, owner, builder, surveyor and project manager I've ever met on any project at the outset really does think that this is the one; this is the project that's going to come in on budget, on schedule and without a hitch. Yep. And there'll be a pink elephant parked permanently in the sky above it.

Optimism is not only endemic in construction, it's epidemic. It's almost as though every professional and tradesman you meet is part of some weird born-again cult: not the Freemasons, more like the Glee Club. They have to be. Building may well be creative, but it is dirty, difficult, wet and hard physical work. The whole process of design is about seeing a perfect conceptual idea being slowly compromised as it takes form in the physical world – that's the truth of the matter. So everybody involved has to be unreasonably optimistic to stay sane. Otherwise they'd all be jumping off the scaffolding. Even building engineers can be unseasonably enthusiastic.

But quantity surveyors are the voice of reason and sanity. They will tell you when they think the project will triple in cost, where the weak moments in the build will come and what to look out for. Not that they are always right because no one can predict what will happen.

You may be intending to be your own builder, your own quantity surveyor and your own project manager. In which case I can offer no advice other than to write everything down and spend four hours every evening on costings and paperwork. Everything that is formalized about design and construction in Britain is bound up in the presumption that you'll follow a prescribed, well-trodden and actually quite effective route. You'll employ an architect, a number of consultants and, eventually, a builder. It's what the whole industry expects. And you'll find that their fees and their own expectations of what's required of them and when fits a preordained schedule of events that I talk about here.

At the beginning I suggest that you get your architect to carry out a feasibility study. They might do this in conjunction with a planning consultant if there are planning issues with the site. A feasibility study typically includes a conceptual design of your new home from which the likely order of costs can be established. Most quantity surveyors will then be able to prepare the information in a format that can be developed at a later stage to form part of the contract documents. So paying them to do this may well not be money wasted. In fact, money spent investigating a design at the early stages is always well spent because you quickly arrive at an understanding of what all the options might be for your site.

The single most significant factor affecting cost will be size. The more area you build, the higher the cost. There are obviously other factors like choice of materials, the architectural style and construction detailing, problems of getting out of the ground (perhaps the single biggest cause of delays) and building shape (simple, square buildings are easier than complex, curved ones). But size is such an overriding criterion that every building put up ends up being costed by the square metre of usable space.

Or square foot in old money. A magic figure seems to be around £100 per square foot, which, thanks to the miracle of serendipity, works out to be around £1,000 per square metre. So the two figures are easily translatable. Just to give you a rough handle on what you might expect to pay, under a grand a square metre is considered wonderful value, but some highly detailed and complex buildings can work out at up to £1,800 per square metre. And don't be surprised if the price of your job fluctuates through the build. Problems, delays and all the little changes that you know you're going to make are all going to put pressure on the budget.

The slow route to pinning down costs

Once you've got an outline cost from your quantity surveyor or, heavens forbid, from your architect, there are then three stages of project development that accompany the design as it, too, develops:

Stage 1 is obtaining planning permission – a point at which the design becomes pretty well finite, at least in terms of exterior detailing, size, massing and layout.

Stage 2 is obtaining building regulations approval – a point at which the drawings supplied for building regulations can provide more detailed information for your engineer or quantity surveyor to cost more thoroughly.

Stage 3 is the issuing of drawings and specifications to the builders for quoting. This is usually called a tender or construction package and, if comprehensive, will contain detailed construction drawings, say, of finishing and joinery details and all the mechanical and electrical components of the building (the accuracy of the package depends, not on your architect, but on whether you have decided which systems you will be installing, how many sockets you will need in the kitchen, and so on). On the production of this package your quantity surveyor might produce a set of guideline figures against which builders' quotes can be measured.

There's no point going through these stages out of order. If planning permission is refused, or there are onerous planning conditions put on the approval, this can affect the way the building design is developed. So, in terms of getting costs fixed, you need to focus on the minimum requirements at each stage and resist, perhaps, both an extravagant design and extravagant spending on consultants.

Stage 3 is perhaps the trickiest stage of all. For a start, once you have all your permissions in place you'll be itching to start work as soon as possible on site. Resist and pour your energies into working with your architect on the detail. Even if you have to change the design later to fit your budget (this will be almost inevitable), you will at least end up with a finite and detailed set of drawings from which a builder can cost.

How much love goes into a tender package?

The answer is a lot of love and time and effort and heartache. It's at this stage that the conceptual and planning design is worked out in all its detail. It'll involve you and your architect and almost certainly a number of consultants. Expect to be paying for some engineering design, a surveyor's time and a soil engineer (who can save tens of thousands of pounds by analysing the ground stability of your site). Bigger jobs might mean asking an electrical or plumbing engineer to be consulted for a day or so to help finalize details on the drawings. And if a significant part of your build involves specialist construction knowledge, then you need to be prepared to pay for it now. If you're building an oak frame and using a generalist architect, he'll want to work with a consultant from a framing company to help specify and detail the oak joinery.

Don't think you can get away with obtaining all these consultant services for free. Some suppliers of goods may be prepared to help out on the never-never in the belief that you're going to push work their way – but they usually want to see a purchase order first. And if you're promising the job to the oak frame company that supplies the consultant,

well, you've just closed down the opportunity to put that package of the build out to competitive tender.

And don't think that putting together the tender package is quick. You need to be prepared to wait as well as pay for it. This is an expensive part of the project: you'll be spending thousands here and you won't have seen even so much as a digger on site. All the more reason, in my view, to have had a quantity surveyor on board earlier to give some ballpark project figures at the planning and building control stages.

And what lovely things are in a tender package?

There's nothing romantic about packages. They're hard-nosed, full of lists and split up into the following:

A written specification describing in detail all aspects of the project starting from the ground up. This will contain standard specification clauses, which can constitute a lengthy part of the document and sometimes seem over the top in their complexity. But in the event of a dispute, they can form an important reference point. Preliminaries also form an important part of a written specification and these will cover items like a site hut, toilet facilities, temporary driveways, scaffold, and so on. Other information can also be included in the specification, such as door and ironmongery schedules (a schedule is effectively a specification or list), lighting schedules, or a schedule of finishes.

Drawings. These will include 1) survey drawings showing the site or the property to be converted as it is now, prior to commencement of the works; 2) planning drawings that have been used to obtain planning permission; 3) building regulations drawings that have been approved and 4) more detailed construction drawings of things like window jambs reveals, door openings, eaves, ground floor to wall junctions and other delights. There should also be electrical layouts and mechanical services layouts and any drawings that have been produced in conjunction with specific trades: a kitchen company or cabinet-makers, for example.

A standard form of contract. This is usually a booklet setting out terms and conditions of the contract and will include references to the contracting parties, any VAT issues, payment terms (including frequency of payment), insurance, procedures for variations and arrangements for retention (that is, how much money you might withhold during and after the construction/contract period until all the snags and faults in the building have been corrected).

How to woo a builder with a tender package

There are perhaps three straightforward ways of working with a builder. The most unusual is to employ a professional project manager as a contracts manager and pay them a fee for the six or eight months it might take to build your house. They engage subcontractors whom they work with regularly and then pass on all the bills for you to pay directly (I outlined this arrangement in Managing your Project, page 268). This is a financially transparent process but something of a closed shop as far as choice of subbies is concerned. You might even be tempted to dispense with the project manager and try running the project yourself to save money. Don't. Four out of five times I see this happen the build goes well over budget through bad management. And you will have the added frustration of dealing with subcontractors who have no particular loyalty to you. The professional project manager has that advantage over you.

Some people opt for the second way of negotiating with a builder they know and trust from experience. Three cheers for that, I say: trust will underwrite every relationship you have on the project and, frankly, I think I'd be prepared to pay slightly over the odds for knowing that my project was in a safe pair of hands.

The third, most commercial route will be to go out to what's called 'competitive tender'. Up to five building companies might be given the construction/tender package and asked to prepare a price and to state this price within a set period. Don't be alarmed if the prices come back differing wildly, or for that matter if two firms don't get back to you at all. One-off house building is a specialism within construction and for that matter your design may be way outside the comfort zone of a builder's experience – they may overprice a project to compensate for the unknown technical risk they might be taking on. And market forces come into play: firms undercut if they need the work and overprice if they don't or think the project is likely to be fraught.

And do remember that the cheapest firm might in the end work out to be the most expensive. Some builders underquote and then pray for 'extras' – the changes that are made to any design during construction – which they can then overquote on, knowing by that then they've already got the job and got you by the short and curlies. I can only here misquote Ruskin, who said that everything has a value and that in reality there's no such thing as a bargain. In the end you pay for everything one way or another.

Signing contracts

Contracts are extremely useful, if only because they make everything clear and force you to marshal your thoughts. Even an exchange of letters can do this, but a build is a complicated thing. A building contract will set out the basis for payment and the rules for things like VAT, insurance and guidelines for when things go wrong. Most contracts require a contract administrator, effectively your architect or surveyor who will implement it and monitor progress and payment against it, issuing forms and approvals at different stages of the build. These will include monthly valuations, variations (changes to the originally contracted specification) and signed-off declarations of approval of works that might trigger a mortgage company's payments to you or stage payments from you to the builder.

The advantages to this established process are that everyone will be familiar with it and consequently ought to behave themselves; you'll get regular financial reports; and you'll be kept in the picture as to progress. If you think about it, the model is simple: you all agree what needs to be done, you create a schedule for it, broken down into bite-sized chunks, then you complete and pay for each chunk in turn.

Of course, there'll be the Great Unseens – items that nobody thought of or that nobody could put their finger on. The glass might have to come from Germany, taking eight weeks to manufacture and then break on the way, resulting in a three-month delay. That means extra costs for you in terms of site insurance, site hut and toilet, scaffolding, and so on.

Or the ground turns out to consist of 36m of soggy clay on top of another 20m of bog. None of which you knew because you didn't want the extra expense of a soil engineer back in the early design stages. Or the price of steel might suddenly jump.

There are also those items that, come the tender process, just might not have been fully specified, for all kinds of reasons – the main one being that you haven't made up your mind whether to make the third bathroom a walk-in wet room or a boudoir. If the contract documents have been put together properly, there will be provision for such unforeseen items and builders cost these with provisional sums, meaning that no one has a clue how much they'll really cost but here's a guesstimate. The variations, as they are recorded, will omit these provisional sums and add the actual expenditure when it's known.

There are other management 'tools' to help you. For example, a building programme or critical path that is supposed to show you what happens when. In reality, these are only any use if kept up to date on an almost daily basis, because deliveries change and subcontractors' start dates move around. But they're useful for seeing whether a delay has any knock-on effects or not.

Coupled with a building programme, the contractor or administrator should also provide you with a cash-flow forecast. Assuming the contract makes provision for you to pay the contractor at, say, four-week intervals, or on completion of certain phases, you're going to need to find significant sums of money on a regular basis. If you need to rob a bank every month, it's useful to know for how much.

Finding the money

I am no expert on money. There's so little advice I can give you here that you'd be better off reading the 'Money' supplement in your weekend newspaper.

In fact, you'd glean more from the financial section of the *Beano* than you would from me. But I can give you one or two pointers.

First, if you're looking to borrow to fund your build, do remember to be cautious about costs and not optimistic. I cannot remember how many times I've suggested to would-be self-builders that it's better to allow a 20 per cent contingency fund across your build and then find you've got a bit to blow on a nice sofa and kitchen at the end, than push your design to the very limit of your funding, only to end up in deep water.

Second, look anywhere and everywhere for money: on the Internet, your bank, newspapers, self-build magazines, inside old mattresses, anywhere.

Third, exploit the value of what you're building. You may just have bought a field with planning permission that cost the entire GDP of Wales, but remember that the value of the site plus the cost of the build does not equal the value of the finished house. The value of the finished made object is always greater than its component parts, even for a highly bespoke and idiosyncratic design. You can use the project's planned value (particularly if you ask a surveyor or estate agent to estimate the project's final worth) to lever more lending out of your mortgage company, something you can attempt not just at the start but, say, two-thirds of the way through the build. The downside of this is that if the project falls apart and you have to sell an unfinished building, you'll be living in penury for the rest of your days.

The last 17.5 per cent

Even more riveting than mortgages is the subject of Value Added Tax. It is not loved. People like me who run small businesses resent being made to be unpaid tax collectors. What was originally a selective trade tax on relatively luxury goods only is now indistinguishable from the sales tax it replaced. Housing, however, is still one anomalous area where VAT is not universally, or for that matter fairly, applied. In very basic terms, a new house shouldn't attract VAT whereas a conversion of an existing building into a habitable dwelling attracts 5%. If you alter or extend a listed building, that, too, does not attract VAT. However, most works to listed buildings will constitute repairs, which do attract VAT. Since I spend up to £10,000 a year 'repairing' my listed home, I think it scandalous that I should have to pay VAT on the work.

To interpret what I've just written, I should explain that by 'attracts' I mean that you'll receive invoices for work and goods that will contain a VAT component, usually set at 17.5 per cent. In other words, VAT exempt is the opposite of 'attracts VAT'. To help you here Customs and Excise produces a self-build claim pack that allows you to recover any VAT you have paid on materials, for example, on a new build. But remember it's the responsibility of the builder to charge you VAT – as a business over a certain size he has to – and if he's in doubt as to the project's VAT status, he will charge it. Moreover, any VAT that can't be exempted at source, by the builder or by a supplier, needs to be recovered by you from Customs and Excise, something that can be done only at the end of the project. So, even if your project is exempt, you'll still have to factor the VAT into your cash-flow forecasts. Since I may not have explained this in the clearest terms, those nice people at Customs and Excise have a helpful website.

Useful Contacts & Addresses

Architectural Bodies

Architect's Registration Board (ARB)
8 Weymouth Street
London W1W 5BU
T: 020 7580 5861
F: 020 7436 5269
E:info@arb.org.uk
www.arb.org.uk

Architecture & Surveying Institute
Chartered Institute of Building
Englemere
Kings Ride, Ascot
Berkshire SL5 7TP
T: 01344 630700 F: 01344 630 777
E: reception@ciob.org.uk
www.ciob.org.uk

The Royal Institute of British Architects (RIBA)
66 Portland Place
London W1B 1AD
T: 020 7580 5533 F: 020 7255 1541
E: info@inst.riba.org
www.org/go/RIBA/home

Architects for Featured Builds

Bl@st Architects
120 Nithsdale Road
Glasgow G41 5RB
T: 0141 423 2955 F: 0141 423 2955
E:info@blastarc.co.uk

Charles Barclay Architects
74 Josephine Avenue
London SW2 2LA
T: 020 8674 0037 F: 020 8683 9696
E: cba@cbarchitects.co.uk
www.cbarchitects.co.uk

Davies Sutton Architects
Penhevad Studios
Penhevad Street
Grangetown, Cardiff CF11 7LU
T: 029 2066 4455 F: 029 2066 4411
office@davies-sutton.co.uk
www.davies-sutton.co.uk

Fashion Architecture Taste (FAT)
FAT Ltd
Appletree Cottage
116–120 Golden Lane
London EC1Y OTL
T: 020 7251 6735 F: 020 7251 6738
E: fat@fat.co.uk
www.fat.co.uk

Featherstone Associates
25 Links Yard
Spelman Street
London E1 5LX
T: 020 7539 3686 F: 020 7539 3687
E: enquries@featherstone-associates.co.uk
www.featherstone-associates.co.uk

FFBA Ltd – Fountain Flanagan and Briscoe Associates
Royal Victoria House
The Pantiles
Tunbridge Wells
Kent TN2 5TE
T: 01892 521525 F: 01892 510414
architects@ffba.freeserve.co.uk

Future Systems
The Warehouse
20 Victoria Gardens
London W11 3PE
T: 020 7243 7670 F: 020 7243 7690
E:email@future-systems.com
www.future-systems.com

George Baxter Associates
Pooh Corner Studio
St George's Lane
Hurstpierpoint
West Sussex BN6 9QX
T: 01273 834866 F: 01273 834706
E: mail@homedesign-online.co.uk

Gillian Horn, Jeremy Till, Sarah Wigglesworth
10 Stock Orchard Street
London N7 9RW
T: 020 7607 9200

Glas Architects
Unit 4
51 Tanner Street
London SE1 3PL
T: 020 7378 7755 F: 020 7378 7722
E: mail@glasarchitects.co.uk
www.glasarchitects.co.uk

Henning Stummel
6 London Mews
London W2 1HY
T: 020 7262 1778 M: 07973 131882
E: mail@henningstummelarchitects.co.uk
www.henningstummelarchitects.co.uk

Hudson Architects
49–59 Old Street
London EC1V 9HX
T: 020 7490 3411 F: 020 7490 3412
E: info@hudsonarchitects.co.uk
www.hudsonarchitects.co.uk

Jamie Fobert Architects
5 Crescent Row
London EC1Y OSP
T: 020 7553 6560 F: 020 7553 6566
E: mail@jamiefobertarchitects.com
www.jamiefobertarchitects.co.uk

Michaelis Boyd Associates
9B Ladbroke Grove
London W11 3BD
T: 0207 221 1237 F: 0207 221 0130
E: info@michaelisboyd.com
www.michaelisboyd.com

Noel Wright Architects
The Old Coach House
Draymans Way, Alton
Hampshire GU34 1AY
T: 01420 542111
www.noelwrightarchitects.co.uk

O'Donnell + Tuomey
20A Camden Row
Dublin 8, Ireland
T: + 353 1 475 2500 F:+ 353 1 475 1479
info@odonnell-tuomey.ie
www.odonnell-tuomey.ie

OMI Architects
31 Blackfriars Road
Salford
Manchester M3 7AQ
E: info@omiarchitects.com
www.omiarchitects.com

Proctor & Matthews
7 Blue Lion Place
237 Long Lane
London SE1 4PU
T: 020 7378 6695 F: 020 7378 1372
E: info@proctorandmatthews.com
www.proctorandmatthews.com

Project Orange
1st Floor, Morelands
7 Old Street
London SC1V 9HL
T: 020 7689 3456 F: 020 7689 3173
E: mail@projectorange.com
www.projectorange.com

Richard Paxton Architects
15 St George's Mews
London NW1 8XE
T: 020 7586 6161 F: 020 7586 7171
E: mail@rparch.com
www.rparch.com

Robert Dye Associates
Linton House
39–51 Highgate Road
London NW5 1RS
T: 020 7267 9388 F: 020 7267 9345
www.robertdye.com

S333 Achitecture + Urbanism
Overtoom 197
1054 HT Amsterdam, NL
T: +31 (0) 20 412 4194 F: +31 (0) 20 412 4187
E: info@s333.org
www.s333.org

Simon Conder Associates
Nile Street Studios
8 Nile Street
London N1 7RF
T: 020 7251 2144 F: 020 7251 2145
E: sca@simonconder.co.uk
www.simonconder.co.uk

Spacelab UK
Unit 404 Kingswharf
301 Kingsland Road
London E8 4DS
T: 020 7684 5392 F: 0207 684 5393
E: info@spacelabuk.com
www.spacelabuk.com

Spratley: Architects + Designers
13 Station Road
Henley-on-Thames
Oxfordshire RG9 1AT
T: 01491 411277 F: 01491 411383
E: design@spratley.co.uk
www.spratley.co.uk

Studio KAP
Central Chambers
109 Hope Street
Glasgow G2 6LL
T: 0141 564 1247 F: 0141 564 1248
E: mail@studiokap.com
www.studiokap.com

Three Fold Architects
The Office
5 Turville Street
London E2 7HX
T: 020 7729 0619 F: 020 7504 8704,
E: info@threefoldarchitects.com
www.threefoldarchitects.com

The Violin Factory
35A Cornwall Road
London, SE1 8TJ
T: 0207 401 7822 F: 0207 401 7602
E:info@mcdlondon.co.uk
www.mcdlondon.co.uk

Building Organisations

Association of Building Engineers
Lutyens House
Billing Brook Road
Weston Favell
Northampton NN3 8NW
T: 01604 404121 Lo.Call: 0845 126 1058
F: 01604 784220
E: building.engineers@abe.org.uk
www.abe.org.uk

British Interior Designers' Association (BIDA)
3/18 Chelsea Harbour Design Centre
London SW10 0XE
T: 020 7349 0800 F: 020 7349 0500
E: enquiries@bida.org
www.bida.org.uk

The Building Centre Group
T: 09065 161 136
www.buildingcentre.co.uk

Council of Registered Gas Installers (CORGI)
T: 01256 372200
www.corgi-gas.com

Institution of Structural Engineers (IStructE)
T: 020 7235 4535
www.istructe.org.uk

National Home Builders Council (NHBC)
T: 01494 735363
www.nhbc.co.uk

Royal Institution of Chartered Surveyors (RICS)
RICS Contact Centre
Surveyor Court
Westwood Way
Coventry CV4 8JE
T: 0870 333 1600 F: 020 7334 3811
E: contactrics@rics.org
www.rics.org.uk

Oil Firing Technical Association (OFTEC)
Foxwood House
Dobbs Lane
Kesgrave
Ipswich IP5 2QQ
T: 0845 65 85 080 F: 0845 65 85 181
E: enquiries@oftec.org
www.oftec.co.uk

Conservation Societies

Birmingham Conservation Trust
Alpha Tower
Suffolk Street
Queensway
Birmingham B1 1TU
T: 0121 303 2664 F: 0121 303 2584
www.birminghamconservationtrust.org

Buildings at Risk Trust
www.buildingsatrisk.org.uk

Cadw
Welsh Assembly Government
Plas Carew
Unit 5/7 Cefn Coed
Parc Nantgarw
Cardiff CF15 7QQ
T: 01443 33 6000 F: 01443 33 6001
E: Cadw@Wales.gsi.gov.uk
www.cadw.wales.gov.uk

English Heritage
Customer Services Department
PO Box 569
Swindon SN2 2YP
T: 0870 333 1181 F: 01793 414926
E: customers@english-heritage.org.uk
www.english-heritage.org.uk

The Georgian Group
6 Fitzroy Square
London W1T 5DX
T: 087 1750 2936 F: 087 1750 2937
E: office@georgiangroup.org.uk
www.georgiangroup.org.uk

Historic Scotland
T: 0131 668 8600
www.historic-scotland.gov.uk

The National Trust
PO Box 39
Warrington WA5 7WD
T: 0870 458 4000 F: 020 8466 6824
E: enquiries@thenationaltrust.org.uk
www.nationaltrust.org.uk

SAVE Britain's Heritage
70 Cowcross Street
London EC1M 6EJ
T: 020 7253 3500 F: 020 7253 3400
E: save@btinternet.com
www.savebritainsheritage.org

Society for the Protection of Ancient Buildings (SPAB)
37 Spital Square
London E1 6DY
T: 020 7377 1644 F: 020 7247 5296
E: info@spab.org.uk
www.spab.org.uk

Twentieth Century Society
70 Cowcross Street
London EC1M 6EJ
www.c20society.org.uk

Victorian Society
1 Priory Gardens
London W4 1TT
T: 020 8994 1019 F: 020 8747 5899
E: admin@victorian-society.org.uk
www.victorian-society.org.uk

Governmental & Environmental Bodies

Association of Environmentally Conscious Building
PO Box 32
Llandysul SA44 5ZA
T: 0845 4569773
E: (general enquiries) graigoffice@aecb.net
www.aecb.net

The CarbonNeutral Company
Bravington House
2 Bravington Walk
Regent Quarter
King's Cross, London N1 9AF
T: 020 7833 6000 F: 020 7833 6049
E: enquiries@carbonneutral.com
www.carbonneutral.com

Centre for Alternative Technology (CAT)
Machynlleth
Powys SY20 9AZ
T: 01654 705950 F: 01654 702782
www.cat.org.uk

Climate Care
115 Magdalen Road
Oxford
OX4 1RQ
T: 01865 207000 F: 01865 201900
E: mail@climatecare.org

Commission for Architecture and the Built Environment (CABE)
1 Kemble Street
London WC2B 4AN
T: 020 7070 6700 F: 020 7070 6777
www.cabe.org.uk

Department for Communities and Local Government (DCLG)
Eland House
Bressenden Place
London SW1E 5DU
T: 020 7944 4400 F: 020 7944 9645
www.communities.gov.uk

Department for Culture, Media and Sport
www.culture.gov.uk

Department for Environment, Food and Rural Affairs (DEFRA)
Nobel House
17 Smith Square
London SW1P 3JR
T: 08459 33 55 77
E: helpline@defra.gsi.gov.uk
www.defra.gov.uk

Fenestration Self Assessment Scheme (FENSA)
FENSA Limited
44–48 Borough High Street
London SE1 1XB
T: 0870 780 2028 F: 020 7407 8307
E: enquiries@fensa.org.uk
www.fensa.co.uk

Forest Stewardship Council
T: 01686 431916
www.fsc-uk.org

Joint Contracts Tribunal (JCT)
9 Cavendish Place
London W1G 0QD
www.jctltd.co.uk

Land Registry
T: 0870 010 8318
www.landregisteronline.gov.uk

WWF: One Planet Living Campaign
WWF-UK
Panda House
Weyside Park
Godalming
Surrey GU7 1XR
T: 01483 426444 F: 01483 426409
www.wwf.org.uk

Media

Build It Magazine
T: 020 7772 8300 F: 020 7772 8584
www.self-build.co.uk

Grand Designs Magazine
T: 01992 570030 F: 01992 570031
E: info@granddesignsmagazine.com
www.granddesignsmagazine.com

Grand Designs is the official magazine for the TV series and is published 12 times a year.

For discounted subscription offers and delivery to your door, visit www.granddesignsmagazine.com or email subscriptions@granddesignsmagazine.com.

Self-Build & Design Magazine
T: 01283 742950 F: 01283 742957
E: production@sbdonline.co.uk
www.selfbuildanddesign.com

Planning Organisations

Royal Town Planning Institute (RTPI)
41 Botolph Lane
London EC3R 8DL
T: 020 7929 9494 F: 020 7929 9490
E: online@rtpi.org.uk
www.rtpi.org.uk

Speer Dade Planning Consultants
10 Stonepound Road
Hassocks BN6 8PP
T: 01273 843737
www.stonepound.co.uk

Planning consultants carrying out planning appraisals, planning applications and statements, objections, appeals and enforcement.

Stonepound Books
10 Stonepound Road
Hassocks BN6 8PP
T: 01273 842155
www.stonepound.co.uk

Publishers of specialist books on planning permission and development land.

Index

Figures in *italics* indicate captions; main references to buildings are indicated in **bold** type.

Acknowledgements

The publishers would like to thank the following for permission to reproduce photographs in this book:

5 Arnhel de Serra, 8–9 Elizabeth Zeschin/FFBA Architects/Media 10 Syndication, 13 Tyson Sadlo, 16–17 Arnhel de Serra, 19 Edina van der Wyck/Hudson Architects/Media 10 Syndication, 23 Dave Young/Noel Wright/Media 10 Syndication, 26 Jefferson Smith/Charles Barclay Architects/Media 10 Syndication, 29 Alex Ramsay/Redcover.com, 32–3 Richard Davies, 37 Chris Tubbs/Michaelis Boyd Associates/Media 10 Syndication, 40 Countryside Properties plc, 41 Left: Arnhel de Serra, top: Jefferson Smith/Sarah Wigglesworth/Media 10 Syndication, bottom: Edina van der Wyck/Hudson Architects/Media 10 Syndication, right: Arnhel de Serra, 42 Left: Alex Ramsay/Redcover.com, right: Jefferson Smith/arcblue.com, 43 Top left: Mel Yates/Project Orange/Media 10 Syndication, centre: Mark Luscombe-Whyte, right: Arnhel de Serra, 45 Paul Massey/Mainstreamimages, 49 Jan Bitter, 54 Nigel Rigden, 58–9 Neil Turner, 66 Mark Luscombe-Whyte, 69 Dennis Gilbert/VIEW & OMI Architects, 73 Tyson Sadlo, 78 Dave Young/Noel Wright/Media 10 Syndication, 83 Chris Tubbs/Michaelis Boyd Associates/Media 10 Syndication, 85 Dennis Gilbert/VIEW & OMI Architects, 90–1 Dave Young/Noel Wright/Media 10 Syndication, 94–5 Tyson Sadlo, 97 Sue Barr/VIEW & Jamie Fobert Architects, 98–99, 110, 112–113 Edina van der Wyck/Richard Paxton Architects/Media 10 Syndication, 100 Dennis Gilbert/VIEW & Jamie Fobert Architects, 103 Mel Yates/Featherstone Associates/Media 10 Syndication, 104 Jefferson Smith/Sarah Wigglesworth/Media 10 Syndication, 106–107 Luke Caulfield, 108 Chris Tubbs/Robert Dye Associates/Media 10 Syndication, 109 Jefferson Smith/Fat Architects/Media 10 Syndication, 111 Edina van der Wyck/Richard Paxton Architects/Media 10 Syndication; Monty's frieze in his kitchen by Dave Lee at Photographic Interiors, 114–115 Jefferson Smith/Fat Architects/Media 10 Syndication, 116 Top left: Dennis Gilbert/VIEW & Jamie Fobert Architects, top right: Sue Barr/VIEW & Jamie Fobert Architects, bottom left and right: Dennis Gilbert/VIEW & Jamie Fobert Architects, 117 Dennis Gilbert/VIEW & Jamie Fobert Architects, 118–119 Sue Barr/VIEW & Jamie Fobert Architects, 120–121 Mel Yates/Featherstone Associates/Media 10 Syndication, 122–125 Jefferson Smith/Sarah Wigglesworth/Media 10 Syndication, 126–129 Chris Tubbs/Robert Dye Associates/Media 10 Syndication, 131, 133 Jan Bitter, 134 Countryside Properties plc, 135 Jefferson Smith/Spratley Architects/Media 10 Syndication, 136 Hufton & Crow/VIEW & Glas Architects, 138–9 Arnhel de Serra, 141 Mark Luscombe-Whyte-The Interior Archive, 142–143 Jan Bitter, 144–145 Countryside Properties plc, 146–147 Jefferson Smith/Spratley Architects/Media 10 Syndication, 148–149 Hufton & Crow/VIEW & Glas Architects, 150–151 Arnhel de Serra, 152 Top left: Mark Luscombe-Whyte-The Interior Archive, top right and bottom: Dennis Gilbert/VIEW & O'Donnell & Tuomey, 153 Mark Luscombe-Whyte-The Interior Archive, 155 Jefferson Smith/Bl@st Architects/Media 10 Syndication, 157 Jefferson Smith/arcblue.com, 158 Tyson Sadlo, 161 Richard Davies, 162 Dave Young/Noel Wright/Media 10 Syndication, 163 Tyson Sadlo, 164–165 Jefferson Smith/Charles Barclay Architects/Media 10 Syndication, 166–167 Jefferson Smith/ arcblue.com, 168–169 Jefferson Smith/Bl@st Architects/Media 10 Syndication, 170–173 Richard Davies, 174–175 Dave Young/Noel Wright/Media 10 Syndication, 176–177 Jefferson Smith/Charles Barclay Architects/Media 10 Syndication, 179 Jefferson Smith/McDonell Associates/Media 10 Syndication, 180–1 Chris Gascoigne/VIEW & Simon Conder Associates, 183 Jefferson Smith/Media 10 Syndication, 184 Beth Evans/Greg Holstead/Media 10 Syndication, 185 Jefferson Smith/McDonell Associates/Media 10 Syndication, 187 Mel Yates/Project Orange/Media 10 Syndication, 188–9 Mark Luscombe-Whyte, 190 Chris Gascoigne/VIEW & Simon Conder Associates, 192–193 Jefferson Smith/Media 10 Syndication, 194 Beth Evans/Greg Holstead/Media 10 Syndication, 195 Beth Evans/Greg Holstead/Media 10 Syndication; artwork above the bath in bedroom by Reuben Welch, 196–197 Jefferson Smith/McDonell Associates/Media 10 Syndication, 198–199 Mel Yates/Project Orange/Media 10 Syndication, 200–201 Mark Luscombe-Whyte, 203 Alex Ramsay/ Redcover.com, 204 NTPL/Dennis Gilbert, 208 Andrew Montgomery/John Falconer Associates/Media 10 Syndication, 212 NTPL/Dennis Gilbert, 213 Kevin McCloud, 214–5 Neil Turner, 216 Edina van der Wyck-The Interior Archive, 217 Alex Ramsay/ Redcover.com, 218–219 Edina van der Wyck-The Interior Archive, 220–221 Neil Turner, 222–223 Alex Ramsay/ Redcover.com, 224–5 Tyson Sadlo, 227 Chris Tubbs/Michaelis Boyd Associates/Media 10 Syndication, 230–1 Edina van der Wyck/Hudson Architects/Media 10 Syndication, 234 Jefferson Smith/Bl@st Architects/Media 10 Syndication, 237 Jefferson Smith/Sarah Wigglesworth/Media 10 Syndication, 245–246 Neil Turner, 252–3 Henning Stummel Architects, 256–7 Mel Yates/Mole Architects/Media 10 Syndication, 265, 272 Tyson Sadlo